Power from the Appalachians

**Recent Titles in
Contributions in Economics and Economic History**

New Perspectives on Social Class and Socioeconomic Development in the Periphery
Nelson W. Keith and Novella Zett Keith, editors

World Population Trends and Their Impact on Economic Development
Dominick Salvatore, editor

Racism, Sexism, and the World-System
Joan Smith, Jane Collins, Terence K. Hopkins, and Akbar Muhammad, editors

Electricity in Economic Development: The Experience of Northeast Asia
Yoon Hyung Kim and Kirk R. Smith, editors

Liberalization and the Turkish Economy
Tevfik Nas and Mehmet Odekon, editors

Windows on a New World: The Third Industrial Revolution
Joseph Finkelstein, editor

China's Rise to Commercial Maritime Power
Irwin Millard Heine

Radical Institutionalism: Contemporary Voices
William M. Dugger, editor

Marching Orders: The Role of the Military in South Korea's "Economic Miracle," 1961-1971
Jon Huer

Linkage or Bondage: U.S. Economic Relations with the ASEAN Region
Hans H. Indorf and Patrick M. Mayerchak

Revolution in the World-System
Terry Boswell, editor

The Will to Win: A Biography of Ferdinand Eberstadt
Robert C. Perez

POWER FROM THE APPALACHIANS

A Solution to the Northeast's Electricity Problems?

Frank J. Calzonetti, Timothy Allison,
Muhammad A. Choudhry,
Gregory G. Sayre, and Tom S. Witt

CONTRIBUTIONS IN ECONOMICS AND ECONOMIC HISTORY,
NUMBER 89

GREENWOOD PRESS
New York • Westport, Connecticut • London

Library of Congress Cataloging-in-Publication Data

Power from the Appalachians.

(Contributions in economics and economic history, ISSN 0084-9235 ; no. 89)
Includes bibliographies and index.
1. Electric power—Northeastern States—Planning.
2. Electric power-plants—Appalachian Region—Planning.
3. Coal-fired power plants—West Virginia—Planning.
4. Interconnected electric utility systems—Northeastern States—Planning. I. Calzonetti, Frank J. II. Series.
HD9685.U6A1187   1989   333.79'32'0974   88-28399
ISBN 0-313-25797-3 (lib. bdg. : alk. paper)

British Library Cataloguing in Publication Data is available.

Copyright © 1989 by Frank J. Calzonetti, Timothy Allison, Muhammad A. Choudhry, Gregory G. Sayre, and Tom S. Witt

All rights reserved. No portion of this book may be reproduced, by any process or technique, without the express written consent of the publisher.

Library of Congress Catalog Card Number: 88-28399
ISBN: 0-313-25797-3
ISSN: 0084-9235

First published in 1989

Greenwood Press, Inc.
88 Post Road West, Westport, Connecticut 06881

Printed in the United States of America

The paper used in this book complies with the Permanent Paper Standard issued by the National Information Standards Organization (Z39.48-1984).

10 9 8 7 6 5 4 3 2 1

Copyright Acknowledgments

The authors and publisher are grateful to the following for permission to reprint their material:

Extracts and Figure 19 from Consolidation Coal Company, CONSOL Coal Research Utility Coal Quality Power Cost Model, 1987. Courtesy of Consolidation Coal Company.

Figure 15 from ECAR/MAAC Coordinating Group, *ECAR/MAAC Interregional Power Transfer Analysis,* 1985. Courtesy of the ECAR/MAAC Coordinating Group.

Figure 14 and Table 19 from John Reeve, "The Location and Characteristics of Recently Completed DC Transmission Schemes in North America," 1984. Courtesy of John Reeve.

This book is dedicated to Dr. Richard A. Bajura,
Director of the West Virginia University
Energy and Water Research Center.
We appreciate his support and guidance.

Contents

Figures	ix
Tables	xi
Preface	xvii
1. Introduction Frank J. Calzonetti	1
2. Planning the Supply of Electricity Frank J. Calzonetti	9
3. Electricity Supply and Demand Frank J. Calzonetti, Tom S. Witt, and Timothy Allison	19
4. Options for Increasing Electricity Supplies Frank J. Calzonetti	53
5. The Power Trade Alternative Frank J. Calzonetti and Muhammad A. Choudhry	75
6. Economic Impacts of the Electricity Industry in West Virginia Frank J. Calzonetti, Gregory G. Sayre, John Merrifield, and Tom S. Witt	101
7. The Politics of Acid Rain and Canadian Power Imports Gregory G. Sayre and Frank J. Calzonetti	129
8. Siting Energy Facilities Frank J. Calzonetti	145
9. A Cost Comparison: Mine-Mouth versus Load Center Power Plant Locations Frank J. Calzonetti, Timothy Allison, Muhammad A. Choudhry, and Tom Torries	171

10. Conclusions 195
Frank J. Calzonetti

APPENDIX 1: Comparison of Control Counties to Impact Counties 201
APPENDIX 2: Data Used for AEP Coal-by-Wire Estimates 205
APPENDIX 3: Power Costing Analysis 207
APPENDIX 4: WVU Electricity Transmission Model 211
APPENDIX 5: Input Data Used in the CONSOL Coal Research Model 221
APPENDIX 6: Coal Analysis and Quality Parameters 223
APPENDIX 7: Power Plant Performance Parameters 225
APPENDIX 8: Constant Dollar Results for Comparative Cost Analysis 227

Index 233

Figures

	Page
1. Weekly Load Curve for a Hypothetical Utility	12
2. United States Federal Regions	21
3. North American Electric Reliability Regions	22
4. Net Generation by Coal, Petroleum, Gas, Nuclear Power, Hydroelectric Power and Other, 1981-1986	29
5. Sales of Electricity to End-Use Sectors, 1949-1986	34
6. Base Case Scenario - Reserve Margin Forecasts Selected Reliability Councils 1987-1996	44
7. Delay Case Scenario - Reserve Margin Forecasts Selected Reliability Councils 1987-1996	46
8. NERC Case Scenario - Reserve Margin Forecasts Selected Reliability Councils 1987-1996	47
9. Number of Power Plant Units Sited in the United States, 1912-1986	54
10. Coalfields of the Conterminous United States	59
11. Interconnections of the North American Electric Reliability Council	76
12. North American Extra-High-Voltage Lines	78
13. Circuit Model of Two Machines Used for Stability Consideration	80
14. North American DC Schemes	82

x   Figures

15. Interconnected System Response for Ontario Hydro
       to NYPP 1000 Megawatt Schedule. 89

16. Major U.S. - Canadian Transmission Line
       Interconnections, 1985 Northeast United
       States, Ontario, Quebec, and New Brunswick. . 92

17. Coal-Oil Price Ratio, 1969-1985. 96

18. Electric Generation Plants in West Virginia. . . .105

19. Structure of the CONSOL Model. 179

A-4-1. Cost of Transmission Line for 100 Miles at
         Different Voltage Levels. 216

A-4-2. Typical Costs for a DC Converter. 218

A-4-3. Break-Even Distance 219

A-4-4. Variation in Break-Even Distance for
         AC and DC Transmission Lines for
         Different Power Ratings 220

Tables

		Page
1.	Comparison of Selected Operational Data by Type of Electric Utility, 1985	11
2.	Completed Central Station Capacity	25
3.	Net Generation by Electric Utilities by Fuel Type, 1986-1987	28
4.	Generating Capacity and Net Generation by Fuel Type and Reliability Council, 1987	30
5.	Generating Capacity and Net Generation by Fuel Type and Reliability Council, 1996	31
6.	Sales of Electricity to Ultimate Consumers, Average Annual Growth Rates by Region	35
7.	Industry Peak Demand Growth Rates	37
8.	Comparison of the Principal Determinants of U.S. Energy Demand as Projected for 2000	37
9.	Projections of U.S. Energy Demand by Sector for 2000	38
10.	Peak Summer/Winter Demands by NERC Region, 1985 and 1986	39
11.	Annual Net Energy for Load by NERC Region, 1985 and 1986	40
12.	Actual and Projected Peak Demands, Summer MW	41
13.	Changes in Capacity Projections	56

xii   Tables

14. Electrical Capacity and Generation
    in the United States, 1985. 57
15. Recoverable Coal Reserves As of January 1, 1976. . . 60
16. Nuclear Power Reactors Currently Under
    Construction by Expected Year of Entry
    Into Operation. 64
17. Planned Capacity Retirements by Year
    As of December 1985 68
18. Installed Industrial Cogeneration Capacity 70
19. North American HVDC Schemes in Operation
    and Active Consideration. 83
20. Generation, Sales, and Bulk Power Transactions
    by Major Private Electric Utilities,
    1965-1985 86
21. Net U.S. Imports of Canadian Power 93
22. Existing, Planned, and Undeveloped
    Hydroelectric Sites in Canada 94
23. Distribution of West Virginia Coal, 1985 103
24. Coal Consumed at West Virginia Power
    Plants, 1985. 104
25. Installed Capacity and Capacity Factors
    at West Virginia Power Plants 106
26. Electric Utility State and Local Taxes in
    West Virginia 107
27. Economic Impacts of Power Plant Construction
    and Operation in Mason County 116
28. Economic Impacts of Power Plant Construction
    and Operation in Pleasants County 117
29. Economic Impacts of Power Plant Construction
    and Operation in Grant County 118
30. Economic Impacts of Power Plant Construction
    and Operations in Putnam County 119
31. State Average Industrial Electricity Prices. . . .121
32. Comparative Cost Analysis for Fabricated
    Plastics Products (SIC 307), 1986 123
33. Comparative Cost Analysis for Industrial
    Inorganic Chemicals (SIC 281), 1986 124
34. Comparative Cost Analysis for Women's
    Clothing (SIC), 1986. 125

Tables   xiii

35. Importance of Electricity Prices: Regional and Local Search. 126

36. Interstate Emissions Transport 133

37. Emissions of Sulfur Dioxide and Nitrogen Oxide in the U.S., 1980 134

38. Highest Sulfur-Dioxide-Emitting Utility Plants . .136

39. U.S. and Canadian SO_2 and NO_x Emissions, 1980-1986 140

40. West Virginia Air Quality Standards. 148

41. Permits Required for Siting a Coal-Fired Power Plant in West Virginia 150

42. Federal, State, Regional, and Local Contacts Required for a Specific 50-mile Transmission Line Application within One State in ECAR . . .152

43. Information to Be Supplied with an Application to Locate a High-Voltage Transmission Line in Pennsylvania 155

44. Response Rate. 165

45. Attitudes Toward the Present Condition of the Community 165

46. Attitudes Toward the Construction of Recent Power Plants. 165

47. Undesirable Impacts of Power Plants. 166

48. Attitudes Toward the Construction of a Hypothetical Coal-Fired Power Plant 167

49. Attitudes Toward the Construction of a Hypothetical Nuclear Power Plant. 167

50. Midwest Coal-by-Wire Costs of Energy, Scrubbers, and New Capacity 173

51. Evaluation of Midwestern Power (Firm) to the Northeast 174

52. Midwest Coal-by-Wire: First Year Costs. 176

53. 30-Year Levelized Cost of New England Power Alternatives. 176

54. Comparative Cost Results for a New Plant in New Jersey vs. Power from West Virginia. . .182

55. Comparative Cost Results for a New Plant in Connecticut vs. Power from West Virginia . .183

xiv Tables

56. Comparative Cost Results for a New Plant in New Jersey vs. Power from West Virginia: Two-Year Accelerated Construction Schedule for West Virginia Plant 187

57. Comparative Cost Results for a New Plant in Connecticut vs. Power from West Virginia: Two-Year Accelerated Construction Schedule for West Virginia Plant 188

58. Comparative Cost Results for a New Plant in New Jersey vs. Power from West Virginia: Two-Year Accelerated Construction Schedule for West Virginia Plant, Two Year Delayed Schedule for New Jersey Plant 189

59. Comparative Cost Results for a New Plant in Connecticut vs. Power From West Virginia: Two-Year Accelerated Construction Schedule for West Virginia Plant, Two-Year Delayed Schedule for Connecticut Plant. 190

A-1-1. Calibration Period Test of the Mason County Control Group. 201

A-1-2. Calibration Period Test of the Pleasants County Control Group. 202

A-1-3. Calibration Period Test of the Grant County Control Group. 203

A-1-4. Calibration Period Test of the Putnam County Control Group 204

A-4-1. Data Used for Transmission Line Cost. 217

A-5-1. Input Data Used in the CONSOL Coal Research Model. 221

A-6-1. Coal Analysis and Quality Parameters. 223

A-7-1. Power Plant Performance Parameters. 225

A-8-1. Comparative Cost Results for a New Plant in West Jersey vs. Power from West Virginia 227

A-8-2. Comparative Cost Results for a New Plant in Connecticut vs. Power from West Virginia 228

A-8-3. Comparative Cost Results for a New Plant in New Jersey vs. Power from West Virginia: Two-Year Accelerated Construction Schedule for West Virginia Plant 229

A-8-4. Comparative Cost Results for a New Plant in Connecticut vs. Power from West Virginia: Two-Year Accelerated Construction Schedule for West Virginia Plant 230

A-8-5. Comparative Cost Results for a New Plant in New Jersey vs. Power from West Virginia: Two-Year Accelerated Constructed Schedule for West Virginia Plant, Two-Year Delayed Schedule for New Jersey Plant. 231

A-8-6. Comparative Cost Results for a New Plant in Connecticut vs. Power from West Virginia: Two-Year Accelerated Construction Schedule for West Virginia Plant, Two-Year Delayed Schedule for Connecticut Plant 232

Preface

In 1984, the West Virginia University (WVU) Energy and Water Research Center commissioned a group of eight faculty members from five disciplines (economics, electrical engineering, geography, mineral resource economics, and political science) to assess the role of electricity exports on the West Virginia economy and to determine whether efforts should be taken to increase electricity exports. The research quickly expanded beyond the state boundary as electricity trade follows physical laws, and disparities in electricity supply and demand exist at the regional level. Moreover, the organizations involved in the electricity industry operate at the multistate rather than the state level. The outcome was a study that considered the electricity trade from the Appalachians to the Northeast and a range of physical, political, geographical, and economic factors influencing this trade and likely to influence it in the future.

The authors of the book were assisted by other WVU investigators on this project. These included Dr. Thomas Torries, Associate Professor of Mineral Resource Economics; Dr. Robert Q. Hanham, Associate Professor of Geography; Dr. Robert Duval, Associate Professor of Political Science; Dr. Patrick Mann, Professor of Economics; and Dr. Gregory Elmes, Associate Professor of Geography. Also contributing to the book was Dr. John Merrifield, Assistant Professor of Economics at the University of Texas, San Antonio. Consolidation Coal Company of Pittsburgh provided the research team with a great deal of assistance and gave the team access to their utility power costing model. Mr. Carl Fink of CONSOL spent much his time with the team working on the scenarios and cost modeling.

In conducting the study, the research team used the services of an advisory board of experts; made visits to various utilities, congressional staff offices, and state and federal energy offices; and sponsored a high-level meeting of utility transmission planners in Morgantown to discuss the electricity situation in the region. Moreover, the researchers were involved in the National Governors' Association Electricity Transmission Task Force. Much of the information gained through these experiences helped the authors to understand the intricacies of the power trade problem.

We wish to thank many people who provided valuable assistance in this study. We must first acknowledge the support of the West Virginia University Energy and Water Research Center and its director, Dr. Richard Bajura. We also wish to acknowledge Dr. Carl Irvin, Associate Director of the center and the other staff, particularly Trina Karolchik. They were most helpful and supportive at every stage of this project. We also wish to thank the advisory board members for their openness in reviewing our work and their willingness to meet with us in Morgantown. These include Dr. Peter Blair of the U.S. Office of Technology Assessment, Billy Jack Gregg of the West Virginia Public Service Commission, David Grubb of the West Virginia Citizen's Action Group, Douglass Henderson and Arthur Wacaster of the Southern States Energy Board, Dr. Edward Hillsman of Oak Ridge National Laboratory, Donald Rodeheaver of Monongahela Power Company, David Serot of the U.S. Energy Information Administration, Eugene Trisko of Stern Brothers, Inc., and Richard Northup of the Appalachian Power Company.

Many other people and organizations provided assistance. These include Mike Fotos of West Virginia Public Energy Authority, Katherine Forbes of the West Virginia's Governor's Office of Community and Industrial Development, Ned Helme of the Center for Clean Air Policy, Ray Maliszewski of American Electric Power, Peter Stern of Northeast Utilities, John Reilly of the Pennsylvania State Energy Office, Edward Rubin of Carnegie-Mellon University, Richard Bright of General Public Utilities, Michael Teitt of the Maryland Power Plant Siting Program, Carl Cater of Allegheny Power Company, and David Boyce and Cynthia Griffin of the University of Illinois.

Most important was the support of the staff at WVU who helped us organize and type the draft and the final report. Tim Schibik, Ph.D. candidate in economics at WVU, assisted in the analysis of electricity supply and demand trends. Karen Betz, of the Regional Research Institute, provided great assistance in helping to organize the WVU Electricity Transmission Meeting, and Mary Lou Meyer did a marvelous job in typing the draft final report. Additional support was given by the WVU College of Arts and Science which provided word processing support for the final typing of the manuscript and by the WVU Center for Economic Research which provided financial support for graduate students in the final completion of the manuscript. Special thanks are due to Joel Halverson, who drafted all of the figures, and to Sherry Fox, who came through at the end and organized a jumbled manuscript into a coherent book.

Power from the Appalachians

1

Introduction

Frank J. Calzonetti

The United States is witnessing the close of an era of power plant construction unparalleled in magnitude at any place or time in the world. In the past twenty years, the equivalent of 400 new large power plants has been constructed in the United States. These power plants were ordered in response to the high growth rate in electricity demand that occurred throughout the 1960s and early 1970s. Although the electricity demand growth rate declined sharply in the mid-1970s and into the 1980s in response to rising electricity prices and economic recession, power plants that were ordered when electricity demand was more robust are still being completed. No new large conventional power plants have been ordered in recent years, and the nation's growth in generating capacity will level off and perhaps even decline in the future as older facilities are retired.

Twenty years ago, conventional wisdom was that electricity demand was going to continue to grow at a healthy rate and that the only way to satisfy future demand requirements was to build new, and usually larger, power plants. There was very little discussion of conservation, alternatives to centralized power plants, or the possibility that electricity demand patterns could change. This mentality brought disaster to many utilities who ordered power plants that were much more expensive than needed and really were not needed in many cases. Today, conventional wisdom is that the nation has an abundance of generating capacity, that there is no end to the ways that conservation can reduce the need for power plants, and that future facilities, if needed, can be smaller and sited very quickly.

There are some who challenge this view. Recently, the U.S. Department of Energy, in the report <u>Energy Security</u>, expressed concern that additional electricity supplies will be needed over the next decade to meet electricity demand growth and to retire and replace aging plants. The study concludes that: "Under cautious assumptions (2 percent demand growth and a 50-year average plant lifetime), in the year 2000 the Nation will need approximately 100 gigawatts of new electric generating capacity (beyond plants now under construction) to maintain adequate electricity supplies" (U.S. Department of Energy 1987, 139). Also, a need for new generating capacity will exist even though utilities have made efforts to manage loads and increase efficiencies. The most urgent need for additional electricity supplies are in the eastern United States. Utility analyst Peter Navarro and former U.S. Secretary of Energy James R. Schlesinger, among others, maintain that the current regulatory process and high inflation rates have caused utilities to choose a cost-minimization strategy which means deferring expenditures on new plants. This strategy, it is argued, is placing the nation in a dangerous position with regard to future electricity supplies (Schlesinger 1986; Navarro 1985).

There is now publicly expressed concern among utility planners and state officials in the Northeast concerning the adequacy of electricity supply. "To assure adequate supplies of electricity for a robust New England economy," says New England Power Pool (NEPOOL) chairman Samuel Huntington, "two types of action are essential: we must vigorously pursue cost-effective programs to use electricity more efficiently and we must add new supplies." On January 14, 1988, New England used 19,311 megawatts of electricity during the peak hour of 6:00 P.M., a full 10-percent increase over the 1987 winter peak (Public Utilities Fortnightly 1988, 21).

Electricity industry executives have found that it has become increasingly expensive to complete new generating capacity. Utility planners estimate that it will take eight years to add a new coal-fired power plant and up to fourteen years to complete a nuclear facility. Many utility executives were surprised by the elasticity of electricity demand, so that rising costs in completing new capacity led to further declines in electricity sales. Also, many discovered that their demand forecasting models, which easily predicted the continuation of growth throughout the 1960s and early 1970s, were ineffective in predicting the weak growth in the late 1970s and 1980s. Industrial demand has declined sharply in many areas of the country, particularly in the Northeast and the Midwest. As demand growth declined, cogeneration from nonutility-controlled sources has provided additional power to some utility systems. Some utility planners expect demand growth to remain relatively flat, but some growing areas such as New England have been surprised by robust growth in response to a revival of the regional economy.

Electric utilities found themselves operating in a new environment. The build and grow directives that previously led to the success of the industry have become a blueprint for disaster as witnessed in several instances throughout the country. Because of the recent experiences of overinvesting in new capacity, most utility planners are now reluctant to invest in new generating capacity. Currently, reserve margins are very high nationwide, and few systems face immediate problems in meeting electricity demand requirements. It should be recognized, however, that one should be careful not to overstate the importance of reserve margins. Reserve margins do not tell the entire story because a system may have a large amount of older or expensive-to-use generating capacity, which is not best suited to serve base load needs.

Even if electricity industry planners wished to construct new generating capacity, it would be difficult to decide upon the type of facility to build. No utility is seriously considering nuclear power, and no one has ordered a nuclear power plant since 1978. Although oil is now a bargain, few expect oil prices to remain competitive with coal. Those utilities with oil-fired capacity are now increasing the use of these facilities, but no utility has announced intentions of adding new oil-based generating capacity. Natural gas is now inexpensive, and it has always been a clean form of energy, but federal legislation once prohibited its use in new power plants. Natural gas is being considered as an alternative worthy of study. Some utilities are considering new technologies, and those that are smaller and modular in nature, which can be brought on-line more quickly are receiving great attention. Combined cycle coal gasification is one such technology. A utility could build combustion turbines without great financial exposure in a short period of time. As the demand situation became clearer, the utility could expand the plant by adding a coal gasification unit.

Alternatives for meeting demand needs other than by constructing new capacity include conservation, load management, life extension of existing plants, and bulk power purchases. Least-cost planning is growing increasingly popular among public utility commissions. The essence of this is for the utility to consider all alternatives that can possibly serve the expected need and to select the cheapest and most reliable alternative and not simply to focus on building additional generating capacity locally. In this light, bulk power purchases become attractive in many cases.

This book considers bulk power purchases. There has been an increasing interest in bulk power transfers in the United States as inter-regional differences in generation cost and capacity grow. The particular focus here is on the northeastern and midwestern regions of the United States and thus with the transfer of power from the coalfields to the Northeast. This topic is investigated in light of increased Canadian power imports, the continued use of nuclear-powered facilities in a highly populated region of the country, and the concern over acid rain.

Certain utilities in the Northeast have ample electricity supplies to satisfy most immediate needs, but these supplies are stretched in meeting peak loads, and the utilities are concerned about how they will satisfy the demand in the mid to late 1990s and beyond. In addition to continued demand growth, albeit at a lower level than in the 1960s and early 1970s, utilities realize that more of their existing capacity will be retired and plant performance will decline. Also, some utilities have mostly oil-fired capacity which they do not wish to rely upon for base load generation.

Other utilities, mostly in the Midwest and the Appalachians, have excess power and are now selling large blocks of power daily to utilities on the East Coast. This power is generated mostly at coal-fired power plants at the mine mouth or on the rivers in the Ohio Valley. This transaction of "economy energy" from coal-fired facilities in this region is beneficial to both the power purchaser, who does not need to use higher cost facilities, and to the seller, who is able to maintain a higher level of utilization of the capacity and to realize savings to the company and the customers. However, the quantity of transfers is limited since the electricity grid was designed principally to ensure system reliability and not to facilitate interregional bulk power transfers.

It is possible that coal-producing regions will increase their electricity-exporting role. Many areas are importing power to satisfy local needs and are considering increasing their power purchases. New England, which has already experienced shortages, depends upon Canadian and New York power to meet peak load demand. New Jersey and eastern Pennsylvania are also importing power. These regions have been experiencing sustained, unexpected growth in electricity demand, which is forcing utilities to consider adding capacity sooner than had been previously thought necessary. Utilities in the East are purchasing economy energy, which is available as long as an excess supply exists elsewhere, but they are evaluating options to obtain "capacity power," which would serve needs for decades.

Efforts are being undertaken to improve the grid system in the eastern United States, which will allow for a greater west-to-east flow. Current exchanges from the Midwest-Appalachians to the East Coast average about 2,400 megawatts daily, but in the future it is expected that East Coast customers will be able to purchase from 3,200 to 3,600 megawatts daily by completing improvements that are already under way. In order to increase transfers above this level, it will be necessary to construct new transmission lines, probably across Pennsylvania, to upgrade existing transmission lines, or to convert some alternating current (AC) lines to direct current (DC) lines.

Power exports from the coalfields can be increased either by the higher utilization of existing capacity or by the construction of new generating capacity. Both offer economic benefits to coal-producing regions. Operating existing facilities more continuously increases coal consumption, which is traced to mining and mining-related jobs and revenue. However, this option may be available only as long as utilities in the Appalachians and Midwest have excess capacity. Electricity sales increase state tax revenue and reduce electricity rates for consumers served by the exporting utility. Constructing new power plants provides these benefits but will result in construction-related growth as well. Based on surveys conducted by the authors in West Virginia, the nation's leading electricity-exporting state, local residents favor proposals to expand the state's electricity-exporting role. Politicians have seized on the issue given to the depressed local economy.

In his January 1985 State of the State address, West Virginia Governor Moore called for the creation of the West

Virginia Public Energy Authority. On October 1, 1985, Governor Moore and Governor Sununu of New Hampshire sent a letter to the governors of fourteen states inviting them to meet in Washington, D.C., to discuss a project to deliver electricity to New England. The idea was to construct a new transmission line between the two regions that would be capable of carrying 2,000 megawatts of electricity. According to the statement, the electricity price differentials between New Hampshire and West Virginia would allow the line to be built as well as scrubbers to be installed on certain West Virginia power plants. The idea is attractive to West Virginia because it would provide a long-term market for the power produced at the state-owned power plants. New Hampshire would find the project acceptable because it represents an alternative to Canadian and local nuclear power and means a reduction in sulfur dioxide emissions from large coal-field power plants, long criticized as the primary source of acid deposition. According to a statement released at the press conference, the most exciting aspect of the proposed line is the benefit in terms of acid rain cleanup. Because of the large differential between power costs in the two regions, it was said that this power sale could potentially offer the cleanup of several existing power plants as part of the bargain. In effect, the New England states would bear the costs of that cleanup through the purchase of the power. According to the Governors, that cleanup effort would come at no increase in electric rates to either regions' ratepayers. This is because the new power would still be cheaper in New England than existing oil-fired capacity. Critics of the plan said that the cost estimates, both for the construction of a new transmission line and for the power generated at the new plant, were unrealistically low. A background paper delivered for the session stated that the total cost of power delivered to New England from this project would be about 4.7 cents/kilowatt-hour, which would include the cost of the new transmission line and the cost of installing scrubbers on existing plants. It was assumed that the cost of the new line would be about $1 million per mile, including right-of-way costs.

This plan was described more fully in a report released by the Center for Clean Air Policy, an organization formed by several governors concerned about acid rain (Neme 1987). Two configurations are considered: The first involves a line directly to New England; the second provides power to eastern Pennsylvania, New Jersey, and New York in addition to New England. Both schemes involve the construction of a new, DC transmission line from the West Virginia northern panhandle to Connecticut. The plan is driven by price differentials in electricity between New England and the Midwest-Northern Appalachians. Because of this price differential and the expected acid rain control cleanup charges, New England utilities would be able to pay for the construction of an additional line plus the cost of installing scrubbers on an existing power plant in Ohio. The cost of power to New England in 1995 (1987 dollars) is expected to range from 4.83 to 5.82 cents/per kilowatt-hour. In addition, the plan would represent from

20 to 40 percent of the total SO_2 emission reduction required in Ohio and West Virginia under the Proxmire-Simpson bill (Neme 1987, 16).

Governor Moore announced that he intended to build four 300 megawatt fluidized-bed power plants at two sites in northern West Virginia to serve customers in the East. The first plant is scheduled to be completed by late 1992, according to Mike Fotos, the director of the West Virginia Public Energy Authority. The cost of each unit will be from approximately $400 to $500 million; the final cost of the entire project will range from $2.5 billion to $3 billion. Allegheny Power System will provide the authority with access to its transmission facilities to move power to the East, and both Allegheny Power and Appalachian Power have offered to assist the state by contracting to operate the plants.

This plan was criticized by some utility representatives. The vice president of the Appalachian Power Company, a subsidiary of American Electric Power, argued that a project of this type would be too expensive for the state economy and that it might represent the largest project ever undertaken by the state. He argued that these new power plants would be unable to compete with the existing power plants owned by the electric utility industry (Charleston Gazette, 1985). American Electric Power leads the nation among the private electric utilities in the amount of power generated for resale.

Recognizing that transmission limitations across Pennsylvania and New York were preventing a greater level of electricity exports, Governor Moore joined with Governor Thompson of Illinois to study the electricity transmission system. Their effort resulted in the formation of a National Governors' Association task force on electricity transmission, which investigated transmission siting and electricity wheeling issues. A report, Moving Power, which was released in February 1987, describes the nature of the nation's transmission system and identifies policy options that could increase interregional electricity transfers (National Governors' Association 1987). Policy recommendations agreed to later listed ways to facilitate the construction of new transmission lines, to resolve disputes among states concerning transmission projects, and to encourage electricity wheeling. Recognition of the transmission limitations and an agreement among the states that ways should be developed to deal with the problem are important first steps in accomplishing the governors' goal.

Building upon the governors' task force, the New England Governors' Conference held a high-level meeting in New Hampshire in March 1988 to consider the coal-by-wire option in meeting New England's power needs. Electric utility presidents, vice presidents, and state government officials representing every state from Maine to West Virginia and Ohio seriously considered ways to bring power into New England from western Pennsylvania, Ohio, and West Virginia. No consensus had been reached by the end of the two-day meeting, except that the concept should be explored further.

Plans to move power from the Midwest and Appalachia to New England call for the construction of new transmission lines across Pennsylvania. Western Pennsylvania coal-fired power plants could also be used to provide power, and substantial wheeling charges could be collected. In November 1987, Pennsylvania Lieutenant Governor Singel, Chairman of the Pennsylvania Energy Office, announced proposed legislation to create the Electric Power Transmission Task Force (Pennsylvania Lieutenant Governor's Office 1987). The task force would include all of the major electric utilities as well as representatives from other interests to "study the potential net economic benefits resulting from changes, including the construction of electric transmission capacity inside and outside of Pennsylvania." The task force was to have two years to complete a study on transmission, but because of political resistance the study has not been initiated.

This book evaluates the question of interregional power trade from the Midwest and Appalachians to the Northeast. First, a background on electricity planning practices is provided, then, details of electricity supply and demand requirements in the Northeast and Appalachians are summarized. Alternatives for meeting demand requirements are discussed. Electricity trade is only one of many ways that can be used to meet anticipated needs. Consideration is then given to the reasons why many in the Midwest and Appalachians are eager to promote the sale of electricity from their region to consumers in other states. This question is examined with respect to the state of West Virginia, where the authors have completed an economic analysis of power exports on the state and local economies.

The major focus of the book is on the economic, political, and institutional issues of expanding coal-by-wire capability. The authors discuss transmission and power plant siting, reporting their research on public acceptance of energy facilities. The book summarizes transmission constraints and reports the results of the authors' comparative cost analysis of siting coal-fired facilities in the Appalachians, and transmitting power to East Coast destinations, to the cost of transporting Appalachian coal to load center power plants. Finally, the authors offer their outlook on electricity trade in the region to promote regional solutions to regional problems and opportunities.

BIBLIOGRAPHY

Charleston Gazette. May 1, 1985. "State Power Plant Program Disastrous."
National Governor's Association. 1987. Moving Power: Flexibility for the Future. Washington, D.C.: National Governors' Association.
Navarro, Peter. 1985. The Dimming of America. Cambridge, Mass.: Ballinger Publishing Company.
Neme, Chris. 1987. Midwest Coal by Wire: Addressing Regional Acid Rain and Energy Problems. Washington, D.C.: Center for Clean Air Policy.
Pennsylvania Lieutenant Governor's Office. Nov. 24, 1987. Press release.

Public Utilities Fortnightly. 1988. "Need for More Electric Capacity Foreseen in New England." 121 (March 17) 21.

Schlesinger, James R. 1986. "The Long-Run Security of the Energy Supply." In *The Future of Electrical Energy*, edited by Sidney Saltzman and Richard E. Schuler, 32-39. New York: Praeger.

U.S. Department of Energy. 1987. *Energy Security*. Washington, D.C.: U.S. Department of Energy.

2

Planning the Supply of Electricity

Frank J. Calzonetti

Electric utilities are finding it most difficult to use traditional ways to plan future capacity. Electricity demand growth has been more difficult to predict, construction schedules for new capacity have lengthened, there is greater bulk power trade, and new generators are entering the business in response to the Public Utilities Regulatory Policy Act (PURPA 210). There is also great uncertainty regarding electricity generating technology and the price of fossil fuels. Nuclear power has been dealt serious setbacks by Three Mile Island, Chernobyl, and Shoreham; acid rain debate lingers over the future of the coal industry; and efforts are being made to reintroduce gas in utility generators. This chapter summarizes the ways in which electricity supply is being planned, who is involved in the planning decisions, and how changes in the industry are influencing prospects for coal-by-wire.

HISTORICAL DEVELOPMENT OF THE ELECTRICITY INDUSTRY

In September 1882, Thomas Edison's Pearl Street Station went into operation providing light to the Wall Street area of New York City (Foster 1977; Rudolph and Ridley 1986, 28-29). Although electric lighting was introduced earlier in other cities, Edison's plant served the nation's financial center and led the way for the electrification of America. Electric lighting developed rapidly across the United States. By the spring of 1883, Edison's Electric Illuminating Company had over 300 generators in operation, and most cities had electric lighting by 1884 (Rudolph and Ridley 1986, p. 31). By 1890 over 1,000 central station power plants were in operation (Glaeser 1957, p. 55). Most plants served local areas with direct current (DC). It was not long, however, before longer distance electricity transmission was developed. In 1892, Southern California Edison was transmitting 10,000 volts of power to a distance of over twenty-eight miles (Rudolph and Ridley 1986, 36).

Hydroelectric power plants were installed at Appleton, Wisconsin (1882), Portland, Oregon (1884), and Niagara Falls, New York (1896). Westinghouse Electric and Manufacturing Company (Niagara Fall's project) was an early case of locating a power plant at the resource and transmitting electricity to a distant load center. Power generated at Niagara Falls was transmitted to Buffalo, New York, via alternating current (AC) lines, and, thereafter, most systems switched to AC transmission (Schuler 1986, 10).

Improving technology allowed the construction of larger generating stations. In 1901, Hartford Electric Light installed a 2-megawatt steam turbine. By 1910, 35-megawatt plants were being located (Rudolph and Ridley 1986, 36). Power companies were providing service to many overlapping areas because many communities believed that competition between firms would reduce costs and improve service. However, overlapping service territories usually resulted in a duplication of facilities and equipment as well as cumbersome accounting procedures. Also, companies were being consolidated into large utility holding companies.

One of Thomas Edison's key aids and the president of Chicago Edison, Samuel Insull, proposed in 1898 that electric utilities be recognized as natural monopolies and be regulated by state commissions. He argued that the industry should remain private but that the state should fix rates and service standards. In 1907, the National Civic Federation's Commission on Public Ownership published its report on the industry. This report advocated the adoption of Insull's plan (Rudolph and Ridley 1986, 38). States quickly introduced legislation regulating the electric utility industry, and the main elements of state electric utility regulation were put into place. Every state, except Delaware and Texas, had established a public utility commission to regulate utilities (not only electric utilities) in the public interest.

ELECTRICITY INDUSTRY TODAY

Table 1 shows the breakdown of the electricity industry in the United States today. In addition to investor-owned utilities, there are public or state-owned utilities, cooperatives, and federal entities. Although investor-owned utilities constitute less than 10 percent of the total industry, they dominate the electric industry's revenues and sales. Public electric utilities include those owned by municipalities, public power districts, or state government organizations. Most of these utilities distribute rather than generate power. Rural electric cooperatives operate in most states; these also have an important power distribution role. Rural electric cooperatives have access to capital through the Rural Electric Administration, the National Rural Utilities Cooperative Finance Corporation, the Federal Financing Bank, and the Bank for Cooperatives. Because of this access to capital, many rural electric cooperatives will enter joint projects with investor-owned utilities. The federal government

Table 1 Comparison of Selected Operational Data by Type of Electric Utility, 1985

Type of Electric Utility	Number of Electric Utilities	Number of Electric Utilities (percent)	Revenues from Sales to Ultimate Consumers (percent)	Sales of Electricity to Ultimate Consumers (percent)	Revenues of Sales for Resale (percent)	Sales of Electricity Available for Resale (percent)
Private	239	9	79	76	43	41
Public/State	1966	61	12	15	15	16
Cooperative	958	29	8	7	23	21
Federal	10	1	1	2	19	22
Total	3173	100	100	100	100	100

Source: U.S. Energy Information Administration, Form EIA-861, "Annual Electric Utility Report."

mainly produces wholesale power, which is sold to other companies for distribution. The Tennessee Valley Authority is the largest federally owned power producer (U.S. Energy Information Administration 1987, 4-5).

PLANNING OF SUPPLY

Except for pumped storage hydroelectric facilities, electric utilities are unable to store power to meet variations in demand. Fig 1 illustrates a typical weekly load curve for an electric utility. Demand rises during the weekdays, usually at specific times during the day when consumers use hot water, appliances, air conditioners, heaters, computers, and so on. At night, the demand drops, and levels usually reach their lowest over the weekends (usually on Sunday). In an ideal situation, in which there is no power trade, the utility will use the generating plants with the highest capital costs and lowest operating costs to serve base load demand. These are generally nuclear, coal-fired, or hydroelectric facilities. As the demand rises, plants with higher operating costs are brought into service until the most expensive-to-run plants are called into service.

Utilities also experience variation in loads over the year. Some utilities experience their greatest loads in the winter, others in the summer. In West Virginia and Ohio, most systems experience their peak loads during the winter months. Maximum peaks often occur in the second day of a cold spell, particularly if it occurs in the beginning of a work week. Other systems record greatest demand levels during the summer. In eastern Pennsylvania and New Jersey, for instance, demand is heavy in the summer because of air-conditioning and other uses. In New England, Mondays are particularly troublesome if the temperatures are high in the summer. When people return to

Figure 1
Weekly Load Curve for a Hypothetical Utility

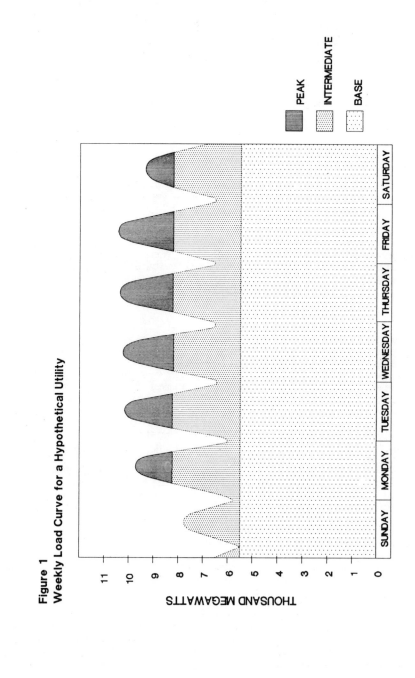

their offices, they turn on air conditioners to lower the temperature in the buildings, which have warmed up over the weekend, as well as computers and other office equipment. Canadian utilities record the highest demand in the winter; summer loads are low because constant air-conditioning is not needed.

Electric utilities need capacity to meet projected peak loads plus capacity in reserve in case of unplanned outages or other problems. The need for a facility may be to satisfy occasional peak loads, or it may be to respond to a steady growth in the base load demand. If a utility experiences problems in meeting peak loads, hydroelectric pumped storage facilities may be favored. If the utility is planning for steady base load growth, then coal-fired and gas-fired power plants are about the only technologies now being considered.

The need for new capacity and the characteristics of this need are determined mainly by the electric utility. However, electric utility demand forecasts are generally evaluated by state public utility commissions to prevent the construction of unneeded additional capacity. It has been argued by some, for instance, that regulated monopolies, such as electric utilities that can add new capacity to their rate base, would build more capital-intensive facilities than they would if they pursued a cost minimization strategy (the Averch-Johnson theory, see Murphy and Soyster 1983). If the utility decides to move ahead with a plan, architecture and engineering contractors are brought in to discuss various power plant concepts. In most cases, architectural and engineering firms will be commissioned to manage the construction of the plant, in close consultation with the electric utility. The primary task at this point is to maintain construction schedules, including the delivery of equipment, so that the new power plant is completed on time. Delays in construction, in permitting, or in delivery of materials escalate the cost of the power plant.

State Public Utility Commissions

The electricity industry is largely regulated by state public utility commissions. The federal government does become involved in matters of interstate commerce and environmental concerns of energy projects, but the state has primary oversight of electric utility operations. Because utilities operate under a state franchise requiring them to provide reliable and economical service to a particular area, many other participants are able to become involved in utility matters.

A franchise is a contract between a private enterprise and a government entity which sometimes provides the enterprise with an exclusive right to provide service in an area (Farris and Sampson 1973, 62). Electric utilities were provided with a franchise to serve a designated area with reliable and economical power. The "mandate to serve" means that if the utility is unable to provide reliable and economical power to the consumers in its service area, the state public utility commission may allow other companies to assume that service district.

Public utility commissions (PUCs) are created by state legislation, but commissioners are often appointed by the governor. Commissions base their decisions on hearings of the available evidence. Public utility commissions require utilities to obtain a "certificate of convenience and necessity," in addition to other charters in order to commence operations. This certificate is needed to expand capacity or service. The utility must demonstrate a need for the service and the ability to provide this service. The PUC will issue the certificate if it is deemed to provide better, more reliable, or more economical service to the customers in the service area. Currently, PUCs regulate over 273 investor-owned utilities which provide almost 80 percent of the nation's electricity (U.S. Energy Information Administration 1987, 4-5).

In return for providing all consumers in the franchise area with reliable service at reasonable prices, PUCs approve rates that give the utilities a "fair and reasonable return" on investment. This fair rate of return compensates the investor-owned utility for the cost of capital investment and provides incentives for future investment (Electricity Policy Project 1983, 2-22). This rate, which has been clarified by several U.S. Supreme Court decisions, is known as the capital attraction standard. That is, the utility must be able to earn enough on its new and existing facilities to attract investors. According to Navarro (1985), many public utility commissions do not provide the utilities with a fair rate of return on investment. He fears that by denying utilities this opportunity, utilities are forced to follow a cost-minimization strategy and are therefore postponing investments in new facilities to minimize losses. Navarro argues that since the early 1970s electric utilities have been hit by rising energy and capital costs not compensated by the rate increases. Thus, electric utilities are failing the capital attraction test, and many have been earning a rate of return below their market rate of capital (Navarro 1985, 10-11). As evidence of this trend, he notes that more utility bonds have been downgraded than upgraded for every year except one since 1973.

Electric utilities are required by law to serve two types of customers: residential, commercial, and industrial customers in the service area; and "requirement customers," the small electric companies that distribute power to other customers but have little of their own generating capacity. Because many requirement customers are dependent upon a single utility, the Federal Energy Regulatory Commission regulates rates for sales to these customers.

Federal Involvement-Regulation

As noted above, the federal government does regulate some electric utility operations. This is largely accomplished through the Federal Energy Regulatory Commission's (FERC) authority over interconnections and wholesale power trade.

The Federal Power Act provides the FERC with authorization to order interconnections between utilities and other entities transmitting or selling electricity (FERC 1981, 55). The FERC has responsibility for regulating both intrastate and interstate power trade, as well as for licensing private hydroelectric facilities. Through sections 205 and 206 of the Federal Power Act, the FERC has certain rate authority concerning interconnected systems and power pools (FERC 1981, 55; Electricity Policy Project 1983, 2-23). The FERC's purpose is to encourage coordination to ensure reliability and efficiency and to minimize discriminatory practices. The FERC's power was expanded somewhat in legislation passed in 1978 which provided a much broader definition of electric utilities to include "any person or state agency which sells electric energy" (16 U.S.C. 796). The new legislation allows a wholesale buyer of electricity to request the FERC to require power wheeling if the supplier ends power sales (FERC 1981, 56-57). The FERC has five commissioners who are appointed by the president and are approved by the Senate.

The prices paid to investor-owned utilities regulated by the FERC for coordination transactions are to be "cost based" unless policy interests demand otherwise (Acton and Bensen 1985, 22-23). In many cases, the price is based on the sellers' incremental costs plus an "adder," either a number of mills per kilowatt-hour or a percentage. In some cases, the prices approved by the FERC have provided the seller with very high percentages added to the incremental cost. This allows the seller to recover costs that may not be directly quantifiable. The FERC has recently made proposals to move toward a more market-based approach to electricity trade. This topic is discussed more fully in Chapter 5.

The federal government is much more heavily involved in nuclear power plant siting and operation. In 1974 the Nuclear Regulatory Commission (NRC) was created to deal with nuclear power plant licensing and regulation. Applicants must submit safety analysis reports, environmental reports, and antitrust information in the early stages of applying for a license to operate a nuclear power plant.

The federal role is much more conspicuous with regard to environmental legislation. Laws and regulations protecting the common environment usually require permits and hearings for power plant construction and many transmission projects. The Federal Environmental Policy Act of 1969 brought federal involvement to a new level with the environmental impact statement requirement for all major projects involving federal action. Although fossil-fuel power plants are not necessarily included, most large projects do require some federal involvement, calling for an environmental impact statement filing. Very often, activities relating to navigable river modification require the Corps of Engineers to assume the lead agency role in the northeastern United States. The corps' responsibility dates to the 1899 Rivers and Harbors Act. The Corps is also concerned with procedures for waste discharges into waterways, although for many power plants this is a very minor factor.

Federal Involvement Policy

Following the 1973-74 oil crisis, the federal government worked vigorously to develop a national energy policy. The immediate effort, "Project Independence," was a plan to achieve near energy self-sufficiency by 1980 by moving to a freer energy market and by encouraging the development of domestic energy resources. The United States enjoyed vast coal reserves and therefore could rely less on oil from the Organization of Petroleum Exporting Countries (OPEC) by using coal in power plants and by developing coal liquefaction and gasification technologies. In 1973, nuclear power accounted for about 5 percent of the nation's electricity, but the U.S. Atomic Energy Commission said that nuclear power's contribution could rise to 22 percent by the early 1980s (DeMarchi 1981, 477). Overall, during the Nixon and Ford administrations, the major effort was to bring forth new energy supplies. The conventional wisdom was that energy demand would continue to grow and that the nation was able through a concerted effort to produce enough energy to meet this demand without relying heavily upon imported oil.

The Carter administration was much more preoccupied with developing a national energy policy. In addition to the development of indigenous energy sources, the Carter plan included a major effort to reduce energy demand. In November 1978, President Carter signed the National Energy Act, which consisted of five components: (1) the National Energy Conservation Policy Act, (2) the Powerplant and Industrial Fuels Use Act, (3) the Public Utilities Regulatory Policy Act, (4) the Energy Tax Act, and (5) the Natural Gas Policy Act. The National Energy Conservation Policy Act encouraged utility conservation programs for residences and other programs designed to reduce energy consumption (Cochrane 1981, 584-85). This was important in stemming the growth in electricity demand. The Powerplant and Industrial Fuels Use Act prohibited the installation of new utility or industrial boilers that used oil or gas. The act also provided funding to attack some of the environmental problems associated with coal use in power plants. The Public Utilities Regulatory Policy Act encouraged cogeneration and provided new rate design standards, including a ban on declining block rates. The Energy Tax Act also aimed to reduce energy demand, and the Natural Gas Policy Act altered pricing schemes to natural gas. President Carter also followed through on earlier initiatives, including the establishment of the Strategic Petroleum Reserve.

The Carter administration, however, was not particularly kind to nuclear power. The Clinch River Liquid Metal Fast Breeder Reactor, which was experiencing expensive cost overruns, was indefinitely deferred by President Carter. The energy policy was generally favorable to the use of coal, but environmental legislation was enacted to control coal mining and use. In particular, the Clean Air Act

Amendments of 1977 required the use of the best available control technology (scrubbers) to reduce power plant emissions. By 1983, 5,700 megawatts of generating capacity were prohibited from burning oil or natural gas and were using coal as their primary fuel (U.S. Department of Energy, Economic Regulatory Administration 1984, 19-20).

CONCLUSION

It is important to recognize the historical mandate of the electricity industry to understand the context in which power trade options are undertaken. Electric utility planners have viewed their primary responsibility to be providing their customers with reliable and economical power. Unless power trade alternatives constitute a means to accomplish these goals, it is doubtful that the industry will support long-term power trade projects. In the next chapter, we discuss electricity supply and demand in the Northeast and demonstrate that there indeed exists a rationale for utility planners to consider long-term interregional power trade alternatives.

BIBLIOGRAPHY

Acton, Jan Paul, and Stanley Benson. 1985. *Regulation, Efficiency, and Competition in the Exchange of Electricity*. Prepared for the U.S. Department of Energy by the Rand Corporation.

Cochrane, James L. 1981. "Carter Energy Policy and the Ninety-fifth Congress." In *Energy Policy in Perspective: Today's Problems, Yesterday's Solutions*, edited by Crawford D. Goodwin et al., 547-600. Washington, D.C.: The Brookings Institution.

DeMarchi, Neil. 1981. "Energy Policy under Nixon: Mainly Putting out Fires." In *Energy Policy in Perspective: Today's Problems, Yesterday's Solutions*, edited by Crawford D. Goodwin et al., 395-473. Washington, D.C.: The Brookings Institution.

Electricity Policy Project. 1983. *The Future of Electric Power in America: Economic Supply for Economic Growth*. Washington, D.C.: U.S. Department of Energy.

Farris, Martin T., and Roy J. Sampson. 1973. *Public Utilities: Regulation, Management, and Ownership*. Boston: Houghton Mifflin Co.

Federal Energy Regulatory Commission. 1981. *Power Pooling in the United States*. Washington, D.C.: Federal Energy Regulatory Commission.

Foster, John Abram. 1979. *The Coming of the Electrical Age to the United States*. New York: Arno Press.

Glaeser, Martin G. 1957. *Public Utilities in American Capitalism*. New York: Macmillan Co.

Murphy, Frederic H., and Allen L. Soyster. 1983. *Economic Behavior of Electric Utilities*. Englewood Cliffs, N.J.: Prentice Hall, Inc.

Navarro, Peter. 1985. *The Dimming of America*. Cambridge, Mass.: Ballinger Publishing Co.

Rudolph, Richard, and Scott Ridley. 1987. *Power Struggle: The Hundred-Year War over Electricity*. New York: Harper and Row Publishers.

Schuler, Richard E. 1986. "Electricity in New York State: The First One Hundred Years as a Prelude to the Future." In *The Future of Electrical Energy: A Regional Perspective of, An Industry in Transition*, edited by S. Saltzman and R. E. Schuler, 7-17. New York: Praeger.

U.S. Department of Energy. 1984. Economic Regulatory Administration. *Powerplant and Industrial Fuels Use Act Annual Report*. Washington, D.C.: U.S. Government Printing Office.

U.S. Energy Information Administration. 1987. *Financial Statistics of Selected Electric Utilities, 1985*. Washington, D.C.: U.S. Energy Information Administration.

3
Electricity Supply and Demand

Frank J. Calzonetti, Tom S. Witt, and Timothy Allison

The future of electric power in the Northeast and Midwest is marked by great uncertainty. Many do not believe that there is anything to worry about at all, that there is ample generating capacity and there exists the potential to bring on new supplies quickly. These new supplies can be in the form of electricity from cogeneration or power imported from Canada. In addition, there is also an abundance of electricity that could be saved by greater conservation efforts.

Others are not convinced that the electricity supply is secure. Many electric power representatives in New England and other rapidly growing areas of the East were not prepared for a return to vigorous rates of electricity demand and now are considering all options to provide power in the future. Although sources of power are available, there is a reluctance on the part of many utility planners to become dependent upon one source of power. In the late 1960s and early 1970s, utility planners responded to marketplace signals and fuel availability by constructing power plants designed to burn imported oil. Oil was cheaper than alternative sources and did not face the new Clean Air Act restrictions that limited the expansion of coal-fired power plants. Almost as soon as these oil-fired facilities were completed, there was an oil boycott and a fourfold increase in oil prices.

In addition to concerns about electricity supply, there is also concern about the price of electricity and its implications for economic growth in the service region. This is particularly true of those electric utilities that now have expensive nuclear-powered facilities that have come on-line. When the Limerick II nuclear power plant, owned by Philadelphia Electric Company, enters the rate base, the price of electricity for residential consumers is expected to rise to almost 15 cents per kilowatt-hour compared to an average of 10.56 cents per kilowatt-hour in 1987 (U.S. Energy Information Administration 1988a). In addition, the recent decision concerning the Shoreham nuclear plant and the resulting planned increased end-user prices will undoubtedly impact economic growth in the Long Island region.

In the meantime, utilities in the Appalachians and Midwest face great uncertainty, not because they face electricity shortages, but because they are not sure how acid rain legislation will affect their coal-fired facilities and the use of coal in new facilities. Generally, utilities in these regions have more than adequate reserve margins and are able and interested in selling power to consuming utilities in the East. The sale of power benefits the utilities by allowing them to operate at a higher utilization rate, which can result in the production of power at a lower cost. These utilities, however, are not in a position to add new capacity for resale for several reasons. First, they cannot be sure that the state public utility commissions will allow new facilities to enter the rate base if they are not needed to satisfy local demand. Second, given the low demand growth rates, they cannot be certain that there will be any change in the need for new capacity in the future. Third, they are selling power because coal-fired facilities are displacing power produced at oil-fired stations. If the price of oil falls, then eastern customers will not demand this power, and, in fact, there could even be an east-to-west flow of electricity. Fourth, acid rain legislation may be passed which could have unknown implications on the economics of coal-fired power plants. Depending upon the nature of the legislation, utilities may be forced to add scrubbers to existing power plants, retire certain units, or resort to other methods to reduce sulfur dioxide and nitrogen dioxide emissions. Finally, Canadian power imports may continue to rise. Because of the price differentials in power produced, it is unlikely that Appalachian and midwestern utilities can compete with Canadian power imports.

There is great uncertainty concerning electricity demand and supply in the Northeast and Midwest. Arguments can be made that the Midwest region has excess capacity and that there is absolutely no need to consider the construction of additional capacity. On the other hand, there is the argument that the Northeast region does face some serious problems that few are willing to face. The remainder of this chapter shows the extent of the uncertainty. We first provide an overview of the electricity demand and supply situation in the region, then summarize the fuel mix, and then show how wrong previous forecasts were. This is followed by an analysis of electricity demand predictions which represent some uncertainty in the future. Finally, both supply and demand are integrated in an analysis of the projections of future reserve margins of various reliability councils under different scenarios.

Throughout the remainder of this book, we summarize data published by the U.S. Energy Information Administration (EIA) in the Department of Energy and data provided by the North American Electric Reliability Council (NERC). Fig 2 shows the U.S. federal regions that are used for tabulating EIA data. Fig 3 is a map of the NERC regions. The NERC was formed by the electricity utility industry in 1968 to promote the reliability and adequacy of bulk power supply

Figure 2
United States Federal Regions

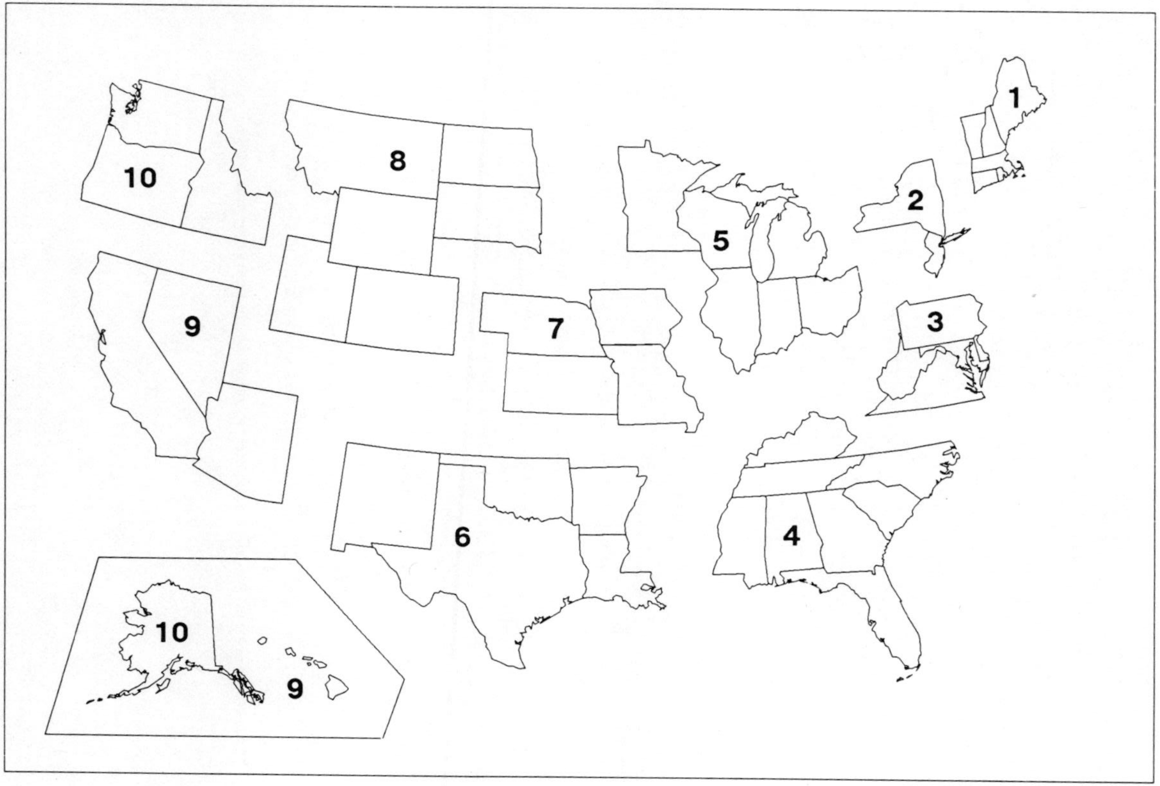

Figure 3
North American Electric Reliability Regions

ECAR–East Central Area Reliability Coordination Agreement
MAIN–Mid-American Interpool Network
MAAC–Mid-Atlantic Area Council
MAPP–Mid-Continent Area Power Pool
NPCC–Northeast Power Coordinating Council
SERC–Southeastern Electric Reliability Council
SPP–Southwest Power Pool
ERCOT–Electric Reliability Council of Texas
WSCC–Western Systems Coordinating Council

Source: Energy Information Administration, <u>Electric Power Annual 1986.</u>

in North America. By reliability is meant the ". . . security of the interconnected transmission network and the avoidance of uncontrolled cascading tripout that may result in widespread power interruptions" (NERC 1984). By adequacy is meant ". . . having sufficient generating capability to meet the electric demands of all customers" (NERC 1984).

CAPACITY AND ELECTRICITY SUPPLY

The distribution of capacity and the choice of power plant vary considerably across the nation. Coal facilities comprise most of the installed capacity in the major coal-producing federal regions: the South Mid-Atlantic; Southeast; Great Lakes; and Rocky Mountain/Northern Great Plains. In addition, coal facilities account for most of the capacity in the Central Plains, not a major coal-producing region, and much of this coal is transported from western mines (EIA 1986). Oil-fired facilities dominate in New England, in the North Mid-Atlantic, and, in the Far West. Gas-fired facilities account for most of the installed capacity of the gas-rich Southwest, and hydroelectric capacity represents the greatest share of capacity in the Northwest. Many of the oil-fired units are also designed to burn natural gas.

Although installed capacity is still increasing, there has been a decline in the completion of new power plants since 1974. There was steady growth in the annual completion of new capacity up to the peak year of 1974, when over 37,000 megawatts of new capacity were added to the national power system. The Southwest was the only region that was building gas-fired power plants in large numbers. A national gas distribution system was not fully in place, and the availability of gas to power plants outside the gas-producing regions was limited. In the Southwest, however, the gas distribution system was in place, and the fuel was plentiful and cheap (Federal Power Commission 1970). The federal government's oil import quotas, designed to increase the strength of the domestic oil-industry, did reduce the availability of oil as a power plant fuel. These quotas were relaxed as oil importing states managed to win permission to purchase foreign oil because of the lack of alternative fuels (Barber 1981). Nuclear power was beginning to emerge as an important source of electricity generation. It must be remembered, however, that it may take twelve years or more to plan and construct a new large power plant, thus new capacity completions reflect past siting decisions.

Factors influencing the choice of power plants include the price of fuels, environmental regulations, electric utility policy, fuel availability, electricity demand, capacity availability, and interrelationships with other systems (EIA 1985c). Numerous econometric studies have investigated factors influencing electricity generation and recent studies show that fuel prices, particularly oil prices, affect the choice of fuel for new generating, capacity (Cowing and Smith 1977; Jaskow and Schmalensee 1983).

Table 2 lists the installed central station capacity completed during selected intervals between 1970 and 1987. The period from 1970 to 1974 was a time when oil import quotas were relaxed, oil was inexpensive, and there was increasing attention to the environmental impacts of energy systems. Also, electricity demand was growing very rapidly. The period from 1975 to 1979 came in the wake of the first surge in oil prices when energy independence was a national goal. Following the 1974 oil price hikes, prices remained stable throughout the remainder of the decade. Since 1980, the nation has experienced the second major jump in oil prices (following the Iranian revolution), then a decline in oil prices, and a long period of low electricity demand growth rates.

The 1970-74 period saw the completion of 141,161 megawatts of new capacity, about 28,232 megawatts per year. Many new power plants were completed in the Southeast and Great Lakes regions. The completion of over 27,176 megawatts of nuclear capacity made nuclear power an important source of electricity in the national picture. It appeared that nuclear power would replace coal as the primary source of electricity during this period. In New England, nuclear power completions exceeded oil-fired completions; nuclear power also made significant gains in the Great Lakes region. A large number of oil-fired units were completed in the South Mid-Atlantic region, a result of the end of the oil import quotas and the declining price of oil. Coal capacity additions were significant in the Great Lakes. In the Southwest, coal-fired facilities were beginning to account for significant additions to the region's capacity, although gas-fired facilities were still the primary choice. Utilities in the region were aware that gas reserves were declining and were showing caution in constructing facilities that needed a reliable source of fuel. Coal-fired additions also occurred in the Far West and Northwest, although oil still dominated in the Far West.

The 1975-1979 period saw 113,875 megawatts of capacity completed, which was 22,775 megawatts per year. The siting activity at this time was a response to the prior period of high electricity growth rates and the belief that growth in electricity demand at historic rates of growth would continue. Seventy percent of the new capacity was added in four regions: the Southeast, the Great Lakes, the Southwest, and the South Mid-Atlantic. The 1970 Clean Air Act greatly reduced the attractiveness of coal in most regions of the country. The 1971 new source performance standards (NSPS) limited emissions of air pollutants from new and modified sources to 1.2 pounds of sulfur dioxide per million British thermal units (Btus) of coal burned and placed limits on particulate and nitrogen oxide emissions. This could be accomplished either by burning low-sulfur coal or by installing expensive scrubbers on new power plants. Most of the nation's low-sulfur steam coal is located in the west, and few utilities in the eastern United States were willing to build coal-fired plants that now required scrubbers. The 1971 NSPS did not take into account the regional variation in coal quality. Utilities could meet the 1.2-pound limit in any way they wished, and

Table 2
Completed Central Station Capacity (by region and period) (net summer capability, MW)

1970-74

Federal Region	Coal	Oil	Gas	Nuclear	Total
New England	0	2,788	0	2,776	5,564
N. Mid-Atlantic	0	7,987	0	1,530	9,517
S. Mid-Atlantic	14,146	5,062	0	4,869	24,076
Southeast	17,523	8,704	1,817	6,106	34,150
Great Lakes	16,657	2,503	232	9,655	29,047
Southwest	2,925	2,018	19,461	902	25,306
Central Plains	5,126	1,101	823	1,338	8,297
Rocky Mts./N. Plains	1,086	233	25	0	1,344
Far West	1,636	4,195	782	0	6,613
Northwest	1,330	867	50	0	2,247
TOTAL	60,429	35,367	23,190	27,176	141,161

1975-79

Federal Region	Coal	Oil	Gas	Nuclear	Total
New England	0	2,698	0	910	3,608
N. Mid-Atlantic	0	2,629	0	3,066	5,695
S. Mid-Atlantic	3,314	3,766	0	4,692	11,772
Southeast	9,209	5,947	1,412	8,365	24,933
Great Lakes	15,248	5,688	50	3,247	24,233
Southwest	10,869	2,116	6,494	0	19,479
Central Plains	6,860	1,510	380	597	9,347
Rocky Mts./N. Plains	6,425	542	0	343	7,310
Far West	2,816	2,100	236	967	6,119
Northwest	51	0	112	1,216	1,379
TOTAL	54,792	26,996	8,684	23,403	113,875

1980-87

Federal Region	Coal	Oil	Gas	Nuclear	Total
New England	0	581	0	1,142	1,726
N. Mid-Atlantic	655	901	0	3,303	4,859
S. Mid-Atlantic	4,461	662	0	4,118	9,241
Southeast	12,978	2,040	27	12,173	27,218
Great Lakes	10,389	107	3	7,909	18,408
Southwest	18,719	73	359	2,152	21,303
Central Plains	6,233	125	278	2,474	9,110
Rocky Mts./N. Plains	8,232	20	21	331	8,604
Far West	2,241	132	248	4,756	7,377
Northwest	508	43	681	1,200	2,433
TOTAL	64,416	4,687	1,617	39,558	110,278

Note: Totals may not equal sum of components due to independent rounding.

Sources: U.S. Energy Information Administration,
1970-1984: Inventory of Power Plants in the U.S.;
1985-1987: Electric Power Annual, p. 4, Table 1.

most of them increased their use of low-sulfur coal. A 500 megawatt power plant burning high-sulfur coal with emission control devices would generate over 91,000 tons of sulfur dioxide a year, compared to the 7,600 tons generated from an identical plant burning low-sulfur coal (Congressional Research Service 1983). Richard Tobin explains that strict state implementation plans reduced the market for high-sulfur coal even further. In states such as Ohio and

Illinois, the quantity of steam coal imported rose dramatically over the period from 1973 to 1981, at the expense of high-sulfur local coal (Tobin 1984). However, the direct substitute of low-sulfur for high-sulfur coal at existing facilities requires the recognition of the relationship between coal characteristics and boiler performance (Coria and Condren 1985). Nuclear facilities, ordered in the 1960s and early 1970s, were rapidly being completed.

The New England and North Mid-Atlantic regions were dominated by oil-fired facilities that had been ordered before the Arab oil embargo, but a large quantity of nuclear capacity also came on-line in the latter region. Coal declined considerably in the South Mid-Atlantic region as nuclear and oil-fired capacity completions dominated. In the Southeast, a major coal-producing region, coal facilities still were dominant, but nuclear power accounted for 33 percent of the total capacity additions. Coal was still the principal fuel in the Great Lakes. In the Rocky Mountains/Northern Great Plains states, where low-sulfur coal was locally available, the growth of coal-fired capacity continued. Coal also grew in importance in the Far West, replacing oil-fired capacity completions.

In the 1980-87 period, capacity additions declined in all regions in the United States, except the Rocky Mountain/Northern Great Plains and the Northwest. Nationally, 110,278 megawatts were completed during this eight-year period. Only 13,784-megawatt capacity was added per year during this time, less than half the rate during the 1970-74 period. There was a slowdown in nuclear completions, and there was a resurgence in the use of coal for power generation, which accounted for 58 percent of all the completed capacity during this period. Part of the completion was affected by the 1978 NSPS, which were designed so that the high-sulfur Appalachian and Central Plains coal-producing regions would not be unfavorably affected. In addition to the emission limit, the 1978 revisions required some removal of the sulfur dioxide in power plants, ranging from 70 percent for power plants burning low-sulfur coal to 90 percent for those burning high-sulfur coal. All new power plants must now employ scrubbers or other control technology. Given that the Carter administration also provided the Powerplant and Industrial Fuels Use Act of 1978, this may have helped the eastern coal fields without sacrificing environmental goals. Although coal facilities accounted for the majority of the new additions, these were added where the fuel was most easily accessible.

Little new capacity was completed in the New England and North Mid-Atlantic regions (both distant from the coalfields) which were facing very high residual oil prices and problems in the completion of nuclear facilities. Capacity completions also declined in the South Mid-Atlantic region, and coal resumed its dominant role. In the Great Lakes, there was also a considerable decline in capacity completions compared to the earlier periods, and coal remained the principal fuel in both regions. A dramatic shift from gas-fired to coal-fired capacity occurred in the Southwest as nearly 88 percent of new capacity in the region was coal fired. Although gas was available at discounted prices, particularly in the summer months, the supply was interruptable because gas use in power plants was set by federal regulations at a lower priority than gas use in the residential and commercial sectors. During 1980 to 1987, the price of gas also increased more in the unregulated intrastate market than in the interstate market so that the gas-producing Southwest experienced rapidly increasing natural gas prices. Utilities with long-term gas supply contracts were cushioned from this somewhat so gas still accounted for a large portion of the total electricity generation. Coal-fired facilities also dominated in the Central Plains and in the Rocky Mountain/Northern Great Plains. The expansion of inexpensive mines in these regions favored the use of coal as a generating fuel. Only in the Southeast did nuclear power increase dramatically relative to coal.

In 1987, electric utilities in the United States generated 2,571,401 gigawatt-hours of electricity, marking a 3.4-percent increase from the previous year (see Table 3). Production of electricity from coal-fired units in 1987 provided 56.9 percent of the total as more electricity was generated from coal than in any previous year. The electric utility industry consumed 718 million tons of coal in that year. Fig 4 suggests that the importance of coal in electricity generation has been increasing, even over the recent period which lacked growth in electricity demand. Nuclear power, second in importance to coal in terms of net generation, provided nearly 18 percent of the total U.S. generation in 1985.

There is great regional variation in the use of fuels for electricity generation in the United States. Table 4 compares the net generation by fuel type and NERC region for the year 1987 with each region's installed generating capacity. The East Central Area Reliability Council (ECAR) generates 92.8 percent of its electricity from coal and only 6.4 percent from nuclear power. The Mid-Atlantic Area Reliability Council (MAAC) is much more dependent upon nuclear power, which provided 34.8 percent of its net generation in 1987. Oil- and gas-fired units (predominately oil-fired units) account for 35.9 percent of the total capacity in the MAAC region but provided only 9.2 percent of the electricity. Coal was far less expensive than oil or gas in 1987. The average cost of coal by use in U.S. electric utilities was 150.6 cents per million Btu compared to 223.4 for natural gas and 297.6 for petroleum (EIA 1988b).

It is expected that coal will maintain its dominance throughout the remainder of this century. Table 5 shows the NERC-forecasted net generation and generating capacity by fuel type for 1996. There will be little change in the proportions of installed generating capacity by 1996 since very few new power plants are expected to be brought on-line in the coming years. A few nuclear units should be completed so that nuclear capacity should account for 15.2 percent of the total installed capacity in the United States. If possible, these nuclear units will be used as base load units so that, by 1996, nuclear power should be

Table 3
Net Generation by Electric Utilities by Fuel Type, 1986-1987 (million kWh)

Fuel Type	1986	1987	Percent change
Coal	1,385,831	1,463,601	5.6%
Petroleum*	136,585	118,480	-13.3%
Gas	248,508	272,618	9.7%
Hydroelectric Power	290,844	249,618	-14.2%
Nuclear Power	414,038	454,818	9.8%
Other**	11,503	12,267	6.6%
TOTAL	2,487,310	2,571,401	3.4%

*Fuel oil, crude oil, kerosene, and petroleum coke
**Includes geothermal, wood, wind, waste, and solar

Note: Totals may not equal sum of components because of independent rounding.

Source: Energy Information Administration, Monthly Energy Review, Dec. 1987, p. 84, Table 7.1.

consistently providing 21.4 percent of the total generation (note that, in 1987, nuclear accounted for 19.2 percent of the net generation). Conditions in the East are not expected to change radically from the current situation. A slight decline in coal generation is expected.

Alternatives to conventional coal, nuclear, and other fossil-fuel plants are not expected to provide a major contribution to electricity supply by the mid-1990s. In the western United States, geothermal and wind energy will be further developed and will account for a growing component of electricity supply.

DEMAND

Critical to the feasibility of a coal-by-wire development strategy are the projected demand for electricity in consuming areas and the projected generation capacity to service these areas. To better understand the economic forces affecting the current and projected electricity demands, this section discusses the factors affecting demand.

A key element in forecasting future trends in electricity capacity and demand is the past. Forecasters have substantially revised their projections of future demand based upon historical experience and, undoubtedly, will continue this revision in the future. Although the demand forecasts have been lowered year after year since the early 1970s there is always the possibility of becoming too

Figure 4
Net Generation by Coal, Petroleum, Gas, Nuclear Power, Hydroelectric Power, and Other, 1981–1986

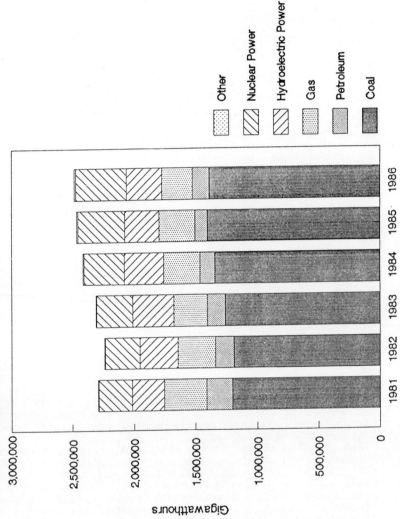

Source: Energy Information Administration, Electric Power Annual, 1981–1986

Table 4
Generating Capacity and Net Generation by Fuel Type and Reliability Council, 1987

	Installed Generating Capacity* (% total)					Electric Energy Production (% total)				
	Nuclear	Coal	Hydro**	Oil/Gas	Other	Nuclear	Coal	Hydro**	Oil/Gas	Other
ECAR	7.1	82.5	0.7	6.0	3.7	6.4	92.8	0.6	0.2	0.0
ERCOT	0.0	29.3	0.6	70.1	0.0	1.1	45.7	0.2	53.0	0.0
MAAC	23.7	35.3	2.4	35.9	2.7	34.8	52.9	2.0	9.2	1.1
MAIN	28.4	56.5	1.2	13.3	0.6	39.3	58.6	1.2	0.6	0.3
MAPP-US	13.0	64.5	10.6	11.8	0.1	21.3	67.1	11.0	0.3	0.3
NPCC-US	17.5	13.5	10.2	52.6	6.2	28.7	18.3	17.1	33.9	2.0
SERC	20.6	48.5	8.2	19.2	3.5	26.3	57.1	5.9	9.5	1.2
SPP	8.9	39.9	3.8	46.7	0.7	15.9	53.5	2.6	27.5	0.5
WSCC-US	8.6	24.4	34.6	26.2	6.2	12.9	34.0	32.1	14.8	6.2
NERC-US	13.9	44.4	10.4	28.0	3.3	19.2	54.8	9.8	14.4	1.8
MAAP-CAN	0.0	29.0	66.2	4.8	0.0	0.0	23.5	75.9	0.6	0.0
WPCC-CAN	20.0	19.4	50.3	10.3	0.0	28.8	10.6	59.8	0.8	0.0
WSCC-CAN	0.0	25.6	61.3	13.1	0.0	0.0	38.5	58.7	2.8	0.0
NERC-CAN	14.2	21.4	53.9	10.5	0.0	20.2	17.6	61.1	1.1	0.0
NERC-MEX	0.0	0.0	0.0	44.3	55.7***	0.0	0.0	0.0	26.5	73.5***
NERC-TTL	13.9	41.7	15.4	3.0	3.0	19.3	50.0	16.3	12.7	1.7

*Summer capabilities, except winter capabilities for Canada
**Conventional hydro. adverse hydro except for a few systems with median or average hydro
***Geothermal capacity and electrical energy production

Source: NERC, 1987 Electricity Supply and Demand 1987-1997, Tables 10 and 17, p. 25-33.

pessimistic in terms of future trends. Of course, these demand forecasts are instrumental in determining the needs for capacity additions in the future. Underlying demand forecasts are assumptions regarding the future trends in factors affecting demand.

Demand forecasts from the NERC are presented and discussed. These forecasts are contrasted with those from the EIA. Although these forecasts are different in many regards, they do provide some basis for determining the future relationship between demand and capacity in relevant markets. Finally, the section presents other national forecasts of demand and contrasts these with the NERC and EIA forecasts.

Theory and Estimation of Electricity Demand

A sizable body of literature has evolved on the theory and estimation of electricity demand (Taylor 1975; Mann and

Table 5
Generating Capacity and Net Generation by Fuel Type and Reliability Council, 1996

	Installed Generating Capacity* (% total)					Electric Energy Production (% total)				
	Nuclear	Coal	Hydro**	Oil/Gas	Other	Nuclear	Coal	Hydro**	Oil/Gas	Other
ECAR	7.4	81.2	0.7	6.1	4.6	9.4	89.8	0.6	0.2	0.0
ERCOT	8.4	33.5	0.4	57.0	0.7	11.4	49.9	0.1	38.2	0.4
MAAC	23.8	33.4	2.2	33.0	7.6	34.6	48.5	1.7	7.9	10.0
MAIN	29.7	55.6	1.2	12.7	0.8	41.8	55.0	1.1	1.3	0.8
MAPP-U.S.	12.5	67.0	9.0	11.4	0.1	18.5	70.2	10.2	0.5	0.6
NPCC-U.S.	21.4	13.3	10.1	46.3	8.9	34.2	18.9	12.6	28.5	5.8
SERC	21.9	45.1	7.3	20.6	5.1	29.7	53.1	5.0	9.7	2.5
SPP	8.9	41.2	3.8	44.8	1.3	15.1	54.1	2.3	27.9	0.6
WSCC-U.S.	8.7	24.2	33.7	23.5	9.9	11.7	34.4	27.9	11.9	14.1
NERC-U.S.	15.2	43.4	10.0	26.2	5.2	21.4	52.9	8.4	12.8	4.5
MAAP-CAN.	0.0	28.0	65.6	6.4	0.0	0.0	27.5	71.8	0.7	0.0
WPCC-CAN.	23.0	18.8	49.3	8.5	0.4	32.3	11.5	54.2	1.6	0.4
WSCC-CAN.	0.0	32.6	55.4	12.0	0.0	0.0	42.3	46.2	7.0	4.5
NERC-CAN.	16.2	22.5	52.1	9.0	0.2	22.5	19.5	54.1	2.6	1.3
NERC-MEX.	0.0	0.0	0.0	52.5	47.5***	0.0	0.0	0.0	35.8	64.2***
NERC-TOTAL	15.3	49.0	15.0	24.2	4.7	21.5	48.1	14.7	11.4	4.3

* Summer capabilities, except winter capabilities for Canada
** Conventional hydro; adverse hydro except for a few systems with median or average hydro
*** Geothermal capacity and electrical energy production

Source: NERC, 1987 Electricity Supply and Demand 1987-1996, Tables 10 and 17, pp. 25 and 33.

Witt 1979; and Bohi 1981 have reviewed this literature). The general theory of commodity demand states that the quantity demanded during a particular time period is a function of the commodity price, other commodities' prices, consumers' incomes, and other variables such as climatic conditions and location. Particular attention is usually placed on the effects of prices and incomes since these factors are determined in the marketplace and vary over time.

Electricity, like other energy products, is not consumed by itself. Electricity demand is interrelated with the demand for the end-use services provided by electricity and the demand for the electricity-utilizing capital stock. The short-run demand for electricity assumes that the electricity-utilizing capital stock is fixed and focuses on the factors which cause changes in the capital stock utilization rate. These factors include price of electricity, income, prices of competing fuels, climatic conditions, and other variables.

The long-run demand for electricity allows the electricity-utilizing capital stock to vary and focuses on the factors which cause changes in the stock, or, alternatively, the choice of energy-using capital stock. The short-run demand for electricity is related to the demand for the services provided by an existing stock; the long-run demand for electricity is related to the demand for the stock. The short-run demand is more inelastic with respect to price and income compared to the long-run demand, since the former does not allow for substitution among capital stocks utilizing different energy sources.

The cross-price elasticity of demand is particularly important in determining the possible substitution among alternative energy sources available to a customer category. The greater the magnitude in this elasticity, the greater the possibilities for substitution of one energy source for another. This is particularly important in generating forecasts of energy consumption since most forecasts are conditional upon a particular scenario regarding energy prices.

Although in theory economists have general expectations regarding the relative magnitudes of short- and long-run price and income elasticities of demand, in practice there is a considerable amount of variation in the magnitude of the estimated values of these elasticities. This variation in the literature can be attributed to the following reasons, among others:

a. Use of different time periods.

b. Use of different observation units, i.e., household, firm, state, utility, and national, among others.

c. Use of static versus dynamic models.

d. Individual equation versus systems of equations model specifications.

e. Econometric methods and models.

f. Measurement of variables, i.e., use of average versus marginal price.

The resulting diversity of demand elasticity estimates can lead to differences in forecasted demand growth since the forecasts are dependent on specific demand forecast scenarios which may not be valid over the long term. After a discussion of historic demand patterns, we turn to the forecasts of future demand for electricity.

Historical Data and Trends

Electricity demand growth has slowed over the past decade compared to the 1960s and early 1970s. Fig 5 provides historical sales of electric energy to ultimate customers in the United States for the period from 1949 to 1986. In recent years, there has been growth over the period in all categories except industrial. For the purposes of this study, data were obtained only for specific regions which are associated with potential markets for Appalachian power.
Table 6 summarizes electricity sales over the period from 1977 to 1986 by ultimate customers for the above-defined regions and states. The magnitude of individual growth rates varies considerably from region to region. Overall, growth during the period ranged from negative for industrial categories in West Virginia to increasing over all customer categories in the Southeast. A comparison of these growth rates with those from the 1960s shows a considerable reduction in growth rates in recent years.

Demand Forecasts

A variety of public and private organizations provides electricity demand forecasts for the entire country, regions, states, and individual utilities. Rather than review all these forecasts, the following section selectively reviews the forecasts that have been widely cited in the literature. In conducting this review, attention is placed on the underlying scenarios and particular assumptions that led to the specific forecast. As in any forecasting exercise, some differences in the forecasts can be attributed to the demand elasticities, forecasting methodologies, data bases, and forecast period scenarios. In the latter case, critical factors are the growth and price of competing energy services and the overall growth in the economy.
National. One source of forecasts for electricity demand over the period from 1987 to 1996 is the NERC.
One of the major activities of the NERC is the compilation of annual forecasts of electricity demand and capacity expansion. Such forecasts are generated at the individual utility level each year and are reported to regional reliability councils for tabulation before submission to the Department of Energy (EP-411 report). Due to the nature of the forecasts, no common scenario is utilized by each utility to ensure consistency of projections. As indicated by the NERC, a number of uncertainties including the following, underlie the projections:

Figure 5
Sales of Electricity to End-Use Sectors, 1949–1986

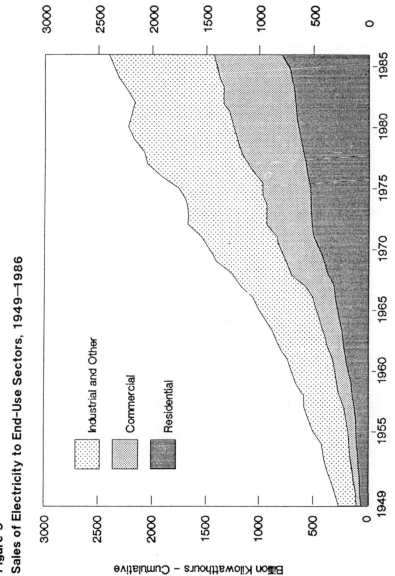

Source: Energy Information Administration, Annual Energy Review 1986.

Table 6
Sales of Electricity to Ultimate Consumers, Average Annual Growth Rates by Region

(1977-1986)

Ultimate Customer Category	Region or State		
	Northeast*	Southeast**	West Virginia
Residential	2.4	3.7	2.3
Commercial	4.2	5.3	4.4
Industrial	0.9	3.2	-1.3
Other	1.7	2.9	2.4
Total	2.5	3.8	1.0

*Maine, New Hampshire, Vermont, Massachusetts, Connecticut, Rhode Island, New York, New Jersey, Delaware, and Maryland.

**Virginia, North Carolina, and South Carolina.

Notes: Totals may not equal sum of components due to rounding. Other sales include public street and highway lighting, other sales to public authorities, sales to railroads and railways, and interdepartmental sales.

Sources: Energy Information Administration, Electric Power Annual, 1981; Tables 124-128; 1986; Table 45.

a. Unanticipated factors affecting future growth of peak demand and energy requirements.

b. The uncertainties associated with the schedules of generating units presently under construction or planned for the future.

c. Unanticipated regulatory restrictions which may affect the operation of existing generating units.

d. Future constraints on fuel availability or use beyond those recognized at the time the projections were prepared (NERC 1984).

Forecasts of the future demand for electricity are released annually by the NERC. These forecasts of electricity demand for the next ten years are aggregations from projections by utility management and are the best indicators available of the demand growth projected by the utilities. Based upon these demand projections, utility management will plan capacity additions and retirements. Table 7 provides the industry peak demand growth rates projected by the NERC from 1977 (Studness 1987, 27; NERC 1988). According to the figures in Table 7, current demand forecasts for the next ten years in NERC-U.S. indicate a projected annual rate of summer peak demand growth of 1.9 percent, which represents a small reduction from the last three projection periods. During the same forecast period (1988-97), the NERC forecasts that the utilities are planning to increase their planned capacity resources (utility and nonutility generating capacity plus scheduled capacity purchases minus capacity sales) by 1.3 percent per year for the period 1988-97 (NERC 1988), which is a reduction from the levels forecasted two years ago (Studness 1987).

Although demand growth has slowed considerably since the 1960s and 1970s, electricity demand will undoubtedly increase over the next ten years. Some analysts suggest that demand growth will parallel the growth in real gross national product (GNP). Substantial variations in GNP growth from year to year will lead to substantial variations in electricity demand. One, however, should not focus on short-run patterns, but on the long-term trends. Such increases will vary among regions and will lead to the need for additional capacity in regions with increasing real GNP.

A wide variety of forecasts from sources other than the NERC are available. These forecasts generally represent national rather than regional forecasts of electricity demand, capacity, and fuel consumption and are predicated on different econometric models and forecast scenarios. Among the organizations with forecasts for the period to 2000, which were available for this study, were Data Resources (DRI), Wharton Econometric Forecasting Associates (WEFA) and the EIA's Annual Energy Outlook (AEO). The principal determinants or economic assumptions underlying each organization's forecasts to 2000 are presented in Table 8. The principal differences in the forecast scenarios involve the higher real GNP growth rate and lower world oil price assumed by the WEFA.

Forecasts of U.S. energy demand by sector for 2000 from the above organizations, based on the determinants in Table 8, are presented in Table 9. Although the data provided in this table relate to all energy resources, it is noted that the projected energy demand growth over the next ten years is primarily due to the growing demand for electric power which is met primarily through the use of coal and nuclear production.

Regional. Table 10 indicates the 1985 and 1986 peak demands by the NERC regions. In the aggregate NERC-U.S., summer peak demand increased by 16,500 megawatts to 477,000 megawatts or an increase of 3.6 percent from 1985 to 1986. On the other hand, winter peak demand fell by 5,800 megawatts to 422,900 megawatts, a decline of 1.4 percent during the same period.

Information on the net energy for load (defined as electrical energy needed to serve utility customers including transmission losses) during 1985 and 1986 by NERC regions is provided in Table 11. In the aggregate NERC-U.S., the net energy for load increased by 32.9 million megawatt-hours to 2532.1 million megawatt-hours, or

Table 7
Industry Peak Demand Growth Rates (Summer Peak).

	NERC-U.S.	
Forecast Published	Forecast Period	Peak Demand Growth Rate (%/year)
1977	1977-86	5.7
1978	1978-87	5.2
1979	1979-88	4.7
1980	1980-89	4.0
1981	1981-90	3.4
1982	1982-91	3.0
1983	1983-92	2.8
1984	1984-93	2.5
1985	1985-94	2.2
1986	1986-95	2.2
1987	1987-96	2.0
1988	1988-97	1.9

Source: NERC, Electric Power Supply and Demand, annual.

Table 8
Comparison of the Principal Determinants of U.S. Energy Demand As Projected for 2000

Determinant	1986 AEO (last years projections)	1987 AEO	DRI	WEFA
Real GNP (billion 1982 dollars)	5,183	5,090	5,142	5,343
Real GNP Growth, 1987-2000 (average annual percent)	2.4	2.2	2.3	2.7
New, High-grade Bond Rate (per cent)	8.7	9.7	9.3	9.6
Industrial Production (index, 1987=1.0)	2.6	2.6	2.8	2.4
World Oil Price (1986 dollars per barrel)	33	30	29	22

Source: Energy Information Administration, Annual Energy Outlook 1986, p. 23, Table 3.

Table 9
Projections of U.S. Energy Demand by Sector for 2000 (quadrillion Btu)

Sector	1986 AEO	1987 AEO	DRI	WEFA
End-Use Consumption				
Residential and Commercial				
Oil and LPG	2.3	2.4	2	NA
Natural Gas	7.3	7.2	6.7	7
Electricity	7.4	7.2	7.1	7
Other*	0.4	0.2	0.4	NA
Total**	17.4	17	16.2	NA
Industrial				
Oil and LPG	8.5	9.4	9.3	NA
Natural Gas	6.7	7.7	5.9	7
Coal	3.1	2.8	3.1	3
Electricity	4.1	4.3	3.6	3.1
Total**	22.5	24.2	21.9	NA
Transportation				
Total**	21	22.3	23.7	NA
Electric Utility				
Oil	3.2	2.8	2.2	1.8
Natural Gas	4	4.8	4.4	2.2
Coal	20.3	19.6	17.8	18.6
Nuclear Power	6.7	6.4	6.4	6.1
Other***	3.9	4.1	4.3	3.9
Total**	38.1	37.6	35.4	32.4
Total End-Use Consumption	60.8	63.4	61.8	NA
Primary Energy Consumption	87.3	89.6	86.5	85.9
Primary Energy/GNP Ratio (thousand Btu per 1982 dollar)	16.8	17.6	16.8	15.7

NA = Not available.

* Includes residential kerosene and steam coal consumption plus commercial kerosene, liquified petroleum gas, and steam coal consumption.
** Excludes renewable resource use in the residential, commercial, and industrial sectors.
*** Includes electricity imports, hydroelectric, geothermal, and other (wood, waste, solar, and wind).

Source: Energy Information Administration, Annual Energy Outlook 1986, p. 24, Table 4.

Table 10
Peak Summer/Winter Demands by NERC Region, 1985 and 1986 (thousands of megawatts)

NERC Region	Summer			Winter		
	1985	1986	Actual Change %	1985	1986	Actual Change %
ECAR	66,293	69,606	5.00	66,667	64,561	1.71
ERCOT	38,062	39,335	3.34	29,776	28,730	3.64
MAAC	37,053	37,564	1.38	31,652	32,807	-3.52
MAIN	32,432	35,943	10.83	29,060	28,036	3.65
MAPP-U.S.	19,936	21,039	5.53	18,803	18,850	-0.25
NPCC-U.S.	40,010	39,026	-2.46	38,233	37,976	0.68
SERC	98,572	105,570	7.10	102,894	101,849	1.03
SPP	45,026	47,123	4.66	34,618	33,877	2.19
WSCC-U.S.	83,119	81,787	-1.60	77,997	76,171	2.40
NERC-U.S.	460,503	476,993	3.58	428,700	422,857	1.38

Source: NERC 1986 Electricity Supply & Demand 1986-1995, 1986, pp. 9-10, Tables 2 and 3.
NERC 1987 Electricity Supply & Demand 1987-1996, 1987, pp. 11-12, Tables 2 and 3.

an increase of 1.3 percent from 1985 to 1986. In comparison, the NERC-Canada net energy for load increased by 3.4 percent to 356.8 million megawatt-hours over the same time period (NERC 1987).

The U.S. peak summer demand projected by the NERC for the period 1988-97 has a 1.9-percent average annual growth rate (see Table 7). The comparable value projected for the period 1987-88 is 2.0 percent (NERC 1987). It is widely recognized that regional growth rates historically have varied considerably among regions and utilities; similarly, one would anticipate similar patterns in demand forecasts. Although the NERC does not provide specific utility forecasts, regional forecasts are available for regional reliability councils. Since this study's focus is on specific regional electricity markets in the eastern United States, attention is directed toward relevant regional reliability councils in this region. Table 12 presents summer peak demand average annual growth rates forecasted for 1987-96 in relevant eastern regional reliability councils.

RESERVE MARGINS

The relative productive position of an electric utility at any future point in time is of primary importance to both consumers and regulators. A common method of analyzing the relative productive position of an electric utility entails the investigation of its reserve margin. Even in its most simple definitional form (one minus the ratio of demand to capacity expressed as a percent), the reserve margin may be used to gain insight into whether a utility's

Table 11
Annual Net Energy for Load by NERC Region, 1985 and 1986 (thousands of megawatt-hours)

NERC Region	1985	1986	Actual Change (%)
ECAR	387,514	390,245	0.70
ERCOT	190,758	190,796	0.02
MAAC	189,671	197,056	3.89
MAIN	173,589	177,190	2.07
MAPP-U.S.	105,797	106,824	0.97
NPCC-U.S.	220,782	227,865	3.21
SERC	522,648	545,409	4.35
SPP	229,677	216,501	-5.74
WSCC-U.S.	478,790	480,218	0.30
NERC-U.S.	2,499,226	2,532,104	1.32

Sources: NERC, 1986 Electricity Supply & Demand, 1986-1995, p. 7, Table 5.
NERC, 1987 Electricity Supply & Demand, 1987-1996, p. 16, Table 5.

generation capacity is sufficient to fulfill its needs. Reserve margins are a key element in this study because of both their simplicity and their widespread acceptance in the industry. For the above reasons, as well as the ability to make different forecast scenarios and comparisons over time, reserve margins are the measures used here.

The analysis presented here focuses on a systematic view patterned after one that could be used by the utilities themselves. After a brief note into the historical patterns surrounding recent reserve margin patterns, the scenarios examined in this study are discussed. These scenarios have been created using statistics gathered by the NERC for the most recent period. The study then uses these statistics to predict the future at various growth rates regarding demand and capacity, assuming utilities either follow their published capacity additions and reductions schedules or delay them one year. Another scenario is then developed by using the NERC's own predictions of demand and capacity. Following the numerical discussion, the limitations to this type of analysis and associated data are discussed.

Table 12
Actual and Projected Peak Demands, Summer MW.
(Midwest and Eastern Regions, (1987-1996 Forecast, in megawatts

	Actual 1986	1987	1988	1989	1990	1991	1992	1993	1994	1995	1996	1987-1996 Ave. Annual Growth (%)
ECAR	69,606	70,056	71,603	73,265	74,340	75,479	76,662	77,926	79,250	80,698	82,081	1.8
MAAC*	37,564	38,000	38,544	39,043	39,481	39,925	40,398	40,869	41,336	41,826	42,355	1.2
MAIN	35,943	36,380	36,948	37,524	38,109	38,726	39,377	40,010	40,631	41,245	41,883	1.6
COM/LTH EDISON	15,100	15,550	15,850	16,150	16,450	16,800	17,150	17,500	17,850	18,200	18,550	2.0
EAST MISSOURI	6,583	6,596	6,678	6,751	6,843	6,936	7,039	7,141	7,224	7,306	7,399	1.3
S CENT ILLINOIS	6,619	6,617	6,719	6,758	6,827	6,883	6,957	7,018	7,086	7,148	7,220	1.0
WIS-N. MICH	7,641	7,617	7,701	7,865	7,989	8,107	8,231	8,351	8,471	8,591	8,714	1.5
NPCC	39,026	41,367	41,931	42,419	42,956	43,648	44,481	45,148	45,847	46,592	47,447	1.5
NEW YORK	23,006	23,360	23,800	24,220	24,700	25,060	25,420	25,720	26,030	26,320	26,620	1.5
NEW ENGLAND	16,020	18,007	18,131	18,199	18,256	18,588	19,061	19,428	19,817	20,272	20,827	1.6
SCRC	105,570	105,770	108,212	111,390	113,892	116,839	119,116	121,337	123,775	126,458	129,054	2.2
FLORIDA	23,270	23,332	23,966	24,515	25,088	25,663	26,220	26,840	27,511	28,320	28,929	2.4
SOUTHERN	27,164	26,311	26,920	27,582	28,050	28,960	29,383	29,995	30,597	31,215	31,837	2.1
TVA	19,980	20,107	20,516	21,550	22,045	22,492	22,862	22,903	23,145	23,493	23,958	2.0
VACAR	35,401	36,020	36,810	37,743	38,709	39,724	40,651	41,599	42,522	43,430	44,330	2.3
NERC TOTAL	524,512	534,463	545,225	557,703	568,827	580,814	592,739	604,094	616,139	628,943	641,203	2.0

*Peak demands are coincident. Contractually interruptable loads are not included.

Source: NERC, 1987 Electricity Supply and Demand 1987-1996, p. 11, Table 2.

Historical Trends

From 1970 through 1984, capacity increased at an annual rate of 4.7 percent, one-third faster than electricity demand (EIA 1986c). This fact lead the EIA to note that reserve margins in some areas had reached 50% in the early 1980s. These numbers were by any standards more than adequate to meet projected demand.

Standards of system reliability, which vary from system to system, usually require reserve margins of between 15 and 20 percent. Such required margins are needed to meet planned outage, scheduled maintenance, and some unexpected system interruptions. Recent data (1985) show that this figure now stands in the range of 30% for the areas contained in this report. History has shown that the electricity industry has chosen a strategy of maintaining status quo in its planning. Current data, as well as some predictions, show that this will lead to a slight upward trend in reserve margins over the next several years. In some instances, however, industry observers are predicting brownouts as early as the summer of 1988 for the Middle Atlantic states (Wall Street Journal, May 24, 1988). It is the results of a continuation of this strategy that this book addresses.

Study Area and Data

The areas chosen for this part of the study are the regional reliability councils; in particular, three specific NERC reliability councils are analyzed in this section: the ECAR, the MAAC, and the Southeastern Electric Reliability Council (SERC). In addition, three subregions are included in order to account for any remaining areas in the eastern United States: the New England subregion (NEPOOL), the New York sub-region (NYPP), and the Virginia-Carolina subregion (VACAR).

The data for these regions were taken from the long-range power supply reports prepared by the NERC (NERC 1987). This data set contains actual values for 1986 along with predictions through 1996. Data were gathered for total peak demand (summer), total generating capability by all fuel types, and planned generating unit additions and subtractions. These figures were then utilized to obtain the reserve margins for the various scenarios.

Scenarios

This section presents the development of the study scenarios representing the various possible future occurrences. The base scenario uses NERC data for 1985 and 1986 to develop future predictions. The delay scenario follows the same pattern allowing for a one-year delay in any capacity alterations. The third scenario uses the NERC predictions to represent the future.

Base Scenario. The data labeled as actual in all of the figures represent data and forecasts developed within this study. This was accomplished by first using the actual

Electricity Supply and Demand   43

peak demand data for 1985 and 1986 and then developing predicted values for 1987-96 from that data. These predicted values were developed for an assumed growth rate in the demand for electricity of 2.4%. It should be noted that this methodology has the capability to develop predictions based on alternative demand growth rates. The figure of 2.4 percent was chosen because it closely approximated the industry-wide expected growth rate at the inception of this research project. Although current NERC forecasts assume a lower growth rate, some recent regional demand growth has been above the forecasted amounts so that the assumed rate of 2.4 percent per year may not be as unrealistic as some might think.

Development of the predicted series was a simple application of a geometric growth rate to the initial actual regional reliability council values for 1985 and 1986. This growth rate was assumed to be constant throughout the period. After demand forecasts were developed, attention was focused on capacity forecasts. As above, only the 1986 actual value from the regional reliability councils was used in the development of the data series.

The actual capacity was adjusted by the planned capacity additions and/or retirements to yield capacity predictions for 1987-96. It should be noted that in all the scenarios reported, the Shoreham nuclear generation facility was not added to the capacity additions. The omission of Shoreham appears to be a good assumption given recent developments between the state of New York and Long Island Lighting Company (LILCO) regarding New York's taking over of Shoreham. These figures were simply an addition to or a subtraction from the actual capacity value based on whether a net capacity addition or deletion was expected during the relevant year. The values used for the capacity changes were also obtained from the regional reliability councils (NERC 1987) and are assumed to be the best information available at that time. Once values for demand and capacity are projected for the period 1987-96, the reserve margin is then calculated. Fig 6 illustrates the values obtained for each of the six study areas.

The data reported in Fig 6 indicate that, after a brief increase in reserve margin, all regional reliability councils are forecasted to show a steady decline after 1987. This is in part due to the lack of planned capacity additions during the forecast time period. This fact is illustrated by data that were gathered by the regional reliability councils from the utilities on their proposed capacity changes. Out of the analyzed regions, VACAR, MAAC, SERC, and NEPOOL are expected to experience critical drops. These areas, without some outside capacity assistance, will be able only to maintain reserve margins of from around 5 to 12 percent by 1996. This projected margin is far below industry-accepted reliability standards of from around 15 to 20 percent, and it is a signal for concern to the industry, regulators, and customers. The other analyzed regions (ECAR and NYPP), although they are also experiencing declines, are not forecast to suffer severe reliability difficulties in this base case scenario. These regions are predicted to

Figure 6
Base Case Scenario – Reserve Margin Forecasts
Selected Reliability Councils 1987–1996

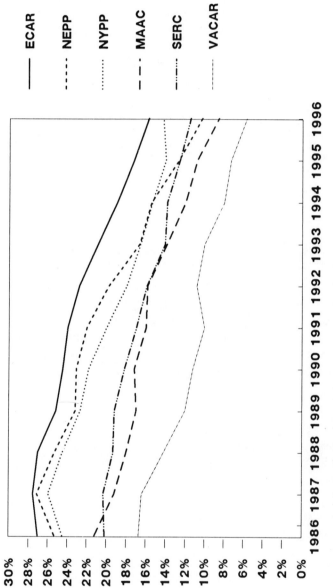

Notes: Electricity demand assumed to grow 2.4 percent per year 1987–1996.
Electricity supply growth based on generation capacity plus planned capacity additions minus planned capacity retirements.

Source: NERC, Electricity Supply and Demand, 1987–1996.

experience declines into the 14-16 percent range by 1994, which is within acceptable limits.

These data show what can happen, given a moderate annual demand growth of 2.4 percent in all regions, adherence to scheduled capacity changes, abandonment of Shoreham, and no outside assistance in the bulk power market. This scenario implicitly assumes that a system is dependent only on its own capacity for reliability. While not very realistic, this assumption is needed to illustrate the potential needs of the various study areas. The main conclusion of this section is that at least four reliability council regions must, in order to remain reliable, either change their future plans or seek outside help for additional electricity supplies.

Delay Scenario. The second scenario builds upon the work of the first. The initial thought was to analyze the impact on the base case scenario of a delay of one year in any firm's capacity changes. Here, the intent was to build in possible alterations in a utility's future plans and to see the effect of these alterations on the first scenario. This approach was extended to all utilities within a given reliability council. The scenario also assumes the exclusion of the Shoreham nuclear plant, as was assumed in the base case scenario.

A delay of one year was applied to both the capacity additions and reductions schedules obtained from the regional reliability councils (NERC 1987). This structure was chosen because of its apparent closeness to the electric utilities' recent history. Most utilities have been forced to delay proposed capacity additions for a wide variety of reasons not discussed in this section of the report. These delays have in turn forced utilities to delay taking other older units off-line in order to maintain reliability. This phenomenon leads to the common delay pattern used above. It should be mentioned that larger lag structures were examined but are not reported here. The results of this delay scenario are presented in Fig 7 for the six study areas. The overall pattern of the predicted values for the reserve margins is similar to the base scenario pattern. The delay scenario consistently leads to predicted reserve margins that are lower than the base scenario.

Given that predicted capacity additions far outweigh capacity retirements, the implication of the above scenario may be a simple caution. The delay of capacity additions, for whatever reason, may have a strong adverse affect on system reliability.

NERC Scenario. This scenario uses the demand and capacity data reported by the NERC in their long-run projections. As in the other two scenarios, it was assumed that Shoreham would be cancelled. The calculations show that reserve margins for these data are slightly more optimistic than in the previous two scenarios (see Fig 8). In this scenario, only the NEPOOL subregion is predicted to have future severe reserve margin problems. Reserve margins for this subregion are predicted to fall into the 4 percent range, which is clearly unacceptable. In addition, this scenario also illustrates the steady decline in the reserve margins of all areas studied.

Figure 7
Delay Case Scenario – Reserve Margin Forecasts
Selected Reliability Councils 1987-1996

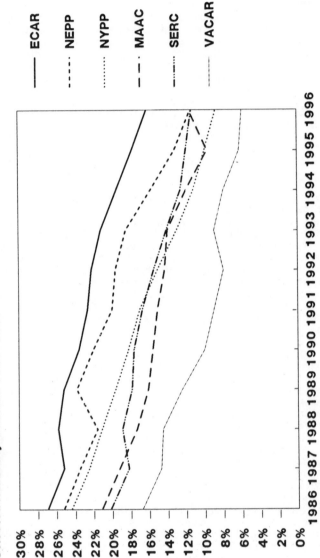

Notes: Electricity demand assumed to grow 2.4 percent per year 1987-1996.
Electricity supply growth based on generation capacity plus planned capacity additions minus planned capacity retirements, both delayed one year.

Source: NERC, Electricity Supply and Demand, 1987-1996.

Figure 8
NERC Case Scenario – Reserve Margin Forecasts
Selected Reliability Councils 1987–1996

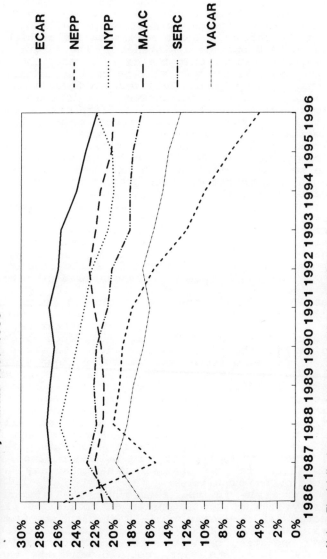

Notes: Electricity demand and supply growth from NERC forecasts.
Source: NERC, Electricity Supply and Demand, 1987–1996.

These conclusions, and this entire scenario, closely match the power requirement study for the West Virginia Public Energy Authority prepared by the Burns and Roe Company (Burns and Roe Company 1985). The Burns and Roe study covered the same study area (along with Florida) and had the same focus as this scenario. The intent was to illustrate through reserve margin analysis which areas are going to need outside assistance in the future to maintain reliability. Given that the data, intent, and conclusions are the same, little more will be said about this scenario.

Scenario Comparison. This section focuses on two areas: the inherent differences between the data used within the various scenarios and the differences between their conclusions. The first and easiest comparison is between the actual and delay scenarios. Here, the data procedure utilized was exactly the same on the demand side. Demand was predicted from 1987 to 1996 by allowing the actual 1986 value to grow geometrically by 2.4 percent. The variation came in through the capacity predictions. Here, the delay scenario pushed all planned additions and retirements back one year. The result was a lower capacity prediction and thus a lower reserve margin. In each of the regions considered, the reserve margin in the delay scenario consistently fell below the one found in the actual scenario. It should be noted that the variation was in most cases minimal near the end of the prediction period due to the small capacity changes predicted to occur at that time.

The NERC scenario must be considered region by region. In three regions, the NERC scenario was consistently much more optimistic than the other two scenarios. In ECAR, MAAC, and NYPP, the NERC scenario predicted higher reserve margins than the other two scenarios. This was due to the lack of optimism for demand growth in these areas shown by the NERC. In the other three regions (SERC, NEPOOL, and VACAR), the NERC scenario predictions were compatible with those of the other two scenarios. Here, the predicted demand figures were, for the most part, the same.

Drawbacks. This section outlines several possible shortcomings involved within the scenarios and their data sets. First, the analysis notes some future factors that will affect the entire industry and its outlook. Next, the focus shifts to some specific areas that may have adverse effects on the results.

A key issue in the future of the electric utility industry is the life extension of coal-fired plants. The expected lifetime of a coal-fired plant is from 30 to 40 years. In 1995, about 60 percent of the existing coal-fired generating capacity will be over thirty years old (EIA 1986c). This fact, coupled with the many factors discouraging new construction, leads to the need for life extension of existing plants.

The concept of plant life extension creates many questions involving system reliability. Data show that forced outages rise sharply after a plant reaches thirty years of age (EIA 1986c). Possible forced outages cause reductions in a system's reliability and thus increase its costs. When coupled with the problem predicted within this section, these plant life extensions may signal even greater reliability problems than anticipated. The capacity figures used in this section assume constant availability and what amounts to constant plant age. The possibility of age-induced outages and other age effects are not foreseen in the data set.

One positive factor that may offset the above age effects to some degree can also be viewed as a drawback to the data set. The notion of technological advances is not built into the data. The data are based on the state of technology that existed at that instant in time from which the data were drawn. It is commonly noted that the possibility of technological advances in the 1990s is great and can only help the predicted situations.

CONCLUSION

The electricity industry has responded to a downturn in electricity demand by postponing decisions to add new generating capacity. The type of capacity now found in the eastern United States is a legacy of power plant capital expenses, fuel prices, and fuel availability anticipated by utility planners who made decisions in the 1960s and early 1970s. Coastal regions are dominated by oil-fired and nuclear facilities whereas the Appalachians and Midwest have more coal-fired facilities. Even with modest electricity demand growth rates, it appears that the NERC's NEPOOL and MAAC will need to add new capacity before the end of this century. Capacity can be added by purchasing power from other systems as well as by constructing new facilities locally. Existing excess capacity in the ECAR region could provide power to the East until new sources of generation are added. In the next chapter, we explore some of the options for increasing electricity supplies.

BIBLIOGRAPHY

Barber, William J. 1981. "The Eisenhower Energy Policy: Reluctant Intervention." In Energy Policy in Perspective: Today's Problems, Yesterday's Solutions, edited by Crawford D. Goodwin, 205-286. Washington, D.C.: The Brookings Institution.
Bohi, Douglas R. 1981. Analyzing Energy Demand Behavior. Baltimore: Resources for the Future.
Burns and Roe Company. 1986. Power Requirements Study for the West Virginia Public Energy Authority.
Congressional Research Service. 1983. An Assessment of the Need for New Electric Capacity. Washington, D.C.: Congressional Research Service.
Coria, Marie R., and Alice E. Condren. 1985. "Which Coal at What Cost." Public Utilities Fortnightly 113:32.
Cowing, Thomas G., and V. Kerry Smith. 1977. "The Estimation of a Production Technology: A Survey of Econometric Analyses of Steam-electric Generation." Land Economics 54:156-86.
Department of Energy. 1983. Energy Projections to the Year 2010. Washington, D.C.: U.S. Department of Energy.

Federal Power Commission. 1970. The 1970 National Power Survey. Parts II and III. Washington, D.C.:U.S. Government Printing Office.

Jaskow, Paul L., and R. Schmalensee. 1983. Markets for Power: An Analysis of Electric Utility Deregulation. Cambridge, Mass.: The MIT Press.

Mann, Patrick C., and Tom S. Witt. 1979. An Economic Analysis of the Electric Utility Sector in the Ohio River Basin Region. Washington, D.C.: Office of Research and Development, U.S. Environmental Protection Agency.

North American Electric Reliability Council. 1988. 1988 Electricity Supply and Demand for 1988-1997. Advance Release. Princeton, N.J.: NERC.

---. 1987. 1987 Electricity Supply and Demand for 1987-1996. Princeton, N.J.: NERC, 1987.

---. 1986. 1986 Electricity Supply and Demand for 1986-1995. Princeton, N.J.: NERC.

---. 1985. 1985 Electric Power Supply and Demand for 1985-1994. Princeton, N.J.: NERC.

---. 1984. 1984 Electric Power Supply and Demand for 1984-1993. Princeton, N.J.: NERC.

Rudolph, Richard, and Scott Ridley. 1986. Power Struggle: The Hundred Year War over Electricity. New York: Harper and Row.

Studness, Charles M. 1987. "NERC's New Ten-Year Forecast of Electric Demand and Capacity." Public Utility Fortnightly, 27-28.

Taylor, Lester D. 1975. "The Demand for Electricity: A Survey." The Bell Journal of Economics and Management Science 6:74-110.

Tobin, Richard J. 1984. "Air Quality and Coal: The U.S. Experience." Energy Policy 12:342-52.

U.S. Energy Information Administration. 1988a. Typical Electric Bills January 1, 1987. Washington, D.C.: U.S. Department of Energy.

---. 1988b. Electric Power Monthly. (January) Washington, D.C.: U.S. Department of Energy.

---. 1987a. Annual Energy Outlook 1986. Washington, D.C.: U.S. Department of Energy.

---. 1987b. Annual Energy Review 1986. Washington, D.C.: U.S. Department of Energy.

---. 1987c. Electric Power Annual 1987. Washington, D.C.: U.S. Department of Energy.

---. 1987d. Monthly Energy Review. Washington, D.C.: U.S. Department of Energy.

---. 1986a. An Analysis of Nuclear Power Plant Construction Costs. Washington, D.C.: U.S. Department of Energy.

---. 1986b. Electric Power Annual 1986. Washington, D.C.: U.S. Department of Energy.

---. 1986b. Annual Energy Outlook 1985. Washington, D.C.: U.S. Department of Energy.

---. 1986c. Annual Outlook for U.S. Electric Power 1986. Washington, D.C.: U.S. Department of Energy.

---. 1986d. Nuclear Power Plant Construction Activity 1985. Washington, D.C.: U.S. Department of Energy.

---. 1985a. Annual Energy Outlook 1984. Washington, D.C.: U.S. Department of Energy.

U.S. Energy Information Administration. 1985b. <u>Coal Distribution</u>. January-December 1984. Washington, D.C.: U.S. Department of Energy.
---. 1985c. <u>Fuel Choice in Steam Electric Generation: Historical Overview</u>. Washington, D.C.: U.S. Department of Energy.
---. 1985d. <u>Electric Power Annual 1985</u>. Washington, D.C.: U.S. Department of Energy.
---. 1984a. <u>Cost and Quality of Fuels for Electric Utility Plants 1983</u>. Washington, D.C.: U.S. Department of Energy.
---. 1984b. <u>Electric Power Annual</u> 1983. Washington, D.C.: U.S. Department of Energy.
---. 1984c. <u>Investor Perceptions of Nuclear Power</u>. Washington, D.C.: U.S. Department of Energy.
---. 1983a. <u>Electric Power Annual</u> 1982. Washington, D.C.: U.S. Department of Energy.
---. 1983b. <u>Nuclear Plant Cancellations: Causes, Costs and Consequences</u>. Washington, D.C.: U.S. Department of Energy.
---. 1983c. <u>Statistics of Privately Owned Electric Utilities, 1981 Annual (Classes A and B Companies)</u>. Washington, D.C.: U.S. Department of Energy.
---. 1982a. <u>Electric Power Annual 1982</u>. Washington, D.C.: U.S. Department of Energy.
---. 1981a. <u>Electric Power Annual 1981</u>. Washington, D.C.: U.S. Department of Energy.
<u>Wall Street Journal</u>. May 24, 1988.

4
Options for Increasing Electricity Supplies

Frank J. Calzonetti

Historically, when an electric utility decided that it needed to increase generating capacity, the major decisions concerned the choice of technology and the location of the power plant. Usually, the utility would locate the facility within the company's service district. In some cases, a utility would enter into a joint purchase agreement with other, usually adjacent utilities to build a larger plant in order to enjoy economies of scale. This formula for responding to electricity demand growth was very successful until the 1970s. In the 1970s, utilities were less successful in predicting the growth in electricity demand; the price of new central station capacity escalated substantially; and siting problems intensified. Utilities now consider a wide menu of alternatives in meeting expected demand requirements, one of which is the construction of new generating capacity. Other options include the retrofitting of existing capacity so that power plants can remain in service well beyond their original date of retirement, load management to reduce the need to use peak power supplies, conservation, and the purchase of power from other systems. They must also consider the role that cogenerated power will have on their need to expand capacity. State public utility commissions now recognize that a better approach in many cases is for the utility to practice least-cost planning so that customers are not burdened with a facility that may be enormously expensive and not fully needed.

NEW CAPACITY

Very few power plants are now under construction in the United States. Fig 9 shows that a rapid expansion of new generating capacity occurred in the United States following World War I, an expansion which continued through most of the 1970s. This new capacity included primarily coal-,

Figure 9
Number of Power Plant Units Sited in the United States, 1912–1986

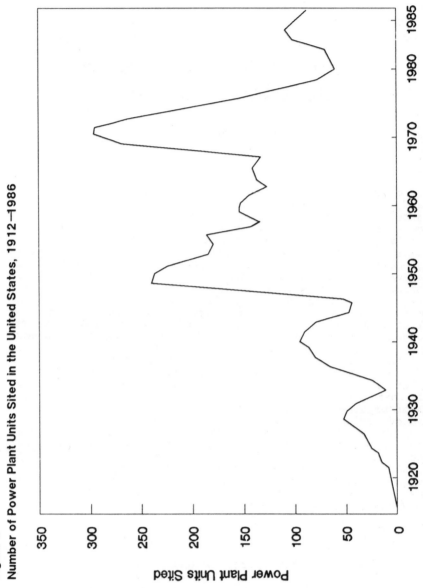

Source: U.S. Energy Information Administration, Generating Unit Reference File and Inventory of Power Plants in the United States, 1983–1986.

Options for Increasing Electricity Supplies   55

oil-, and gas-fired power plants, nuclear stations, and hydroelectric facilities. The development of energy facilities across the nation was uneven because of local resource endowments, cost and availability of fuels, siting constraints, and utility preferences.

Electricity demand declines (beginning in the mid-1970s) led to a slowdown in orders for new generating capacity. It has been only recently, however, that other options have been considered to be viable substitutes for the construction of new power plants. Table 13 illustrates recent trends in power company planning and capacity completions. In 1981, 157,400 megawatts of new generating capacity were scheduled to be completed over the period from 1982 to 1991. As shown in column A, coal facilities were to account for 75,899 megawatts and nuclear power, 69,438 megawatts of this new capacity. A ten-year time period is not considered to be a very distant future horizon for power company planners. Probably considerable work was under way on many of these new power plants. As of December 31, 1985, there were only 72,098 megawatts of projected capacity additions. Electricity planners have been scaling down their projections for the need for new generating capacity.

Over the period from 1982 to 1985, 62,896 megawatts of this new capacity were added, as shown in column B of Table 13. Coal-fired power plants accounted for the largest percentage of these new facilities. Column C lists the remainder of the 1981 projected capacity, and column D shows the total projected capacity recorded in 1985 for the period from 1986 to 1995. Finally, column E lists the total capacity either brought on-line since 1982 or projected to 1995 compared to that projected for the period from 1982 to 1991. According to these figures, 22,406 megawatts of total generating capacity have been cancelled in the United States. Nuclear capacity accounts for the greatest loss with 16,956 megawatts lost; coal is next with 9,018 megawatts lost. According to a survey conducted by *Power Engineering* (Smock and Reynolds 1987), seven nuclear plants began operations in 1986. Of the twenty nuclear power plants now under construction, seventeen have definite target dates. By 1990, nuclear power plant construction will have ended in the United States. The last nuclear power plant was ordered in 1978, but it was cancelled. The last nuclear power plant that has not been cancelled was ordered in 1974 (Smock and Reynolds 1987).

By contrast, it appears that natural gas has experienced renewed growth compared to 1981 electricity company forecasts. Natural gas prices are much lower than previously expected. If the gas is available at a competitive price, utilities find gas to be a clean and convenient fuel. Gas can also be used in modular units which can be placed into operation in a short period of time. Some utilities are finding integrated coal-gasification combined cycle (IGCC) plants to be very attractive. These plants involve the construction of gas turbines, and, as the demand for electricity increases, coal gasifiers can be added as needed (Spenser, Alpert, and Gilman 1986). The 100-megawatts Coolwater IGCC plant has

Table 13
Changes in Capacity Projections

Fuel	A 1981 Projected Capacity (MW)	B Brought On-Line 1982-85	C A-B	D 1985 Projected Capacity	E Capacity Cancelled (-) or Added (+)
Coal	75,899	34,016	41,833	32,815	-9,018
Oil	1,400	373	1,027	1,251	+224
Gas	6	770	(764)	3,230	+3,994
Water	10,657	6,038	4,619	4,019	-600
Nuclear	69,438	21,699	47,739	30,783	-16,956
Total	157,400	62,896	94,504	72,098	-22,406

Source: U.S. Energy Information Administration. 1981-86. *Inventory of Power Plants in the United States*.

been acclaimed as an engineering success, and eastern utilities, namely Virginia Power and Potomac Edison, are planning to develop IGCC facilities. Overall, there is very little commitment to new base load conventional power plants. Work on existing plants is proceeding, but very few utilities are announcing plans to order to power plants.

Coal

The importance of coal to U.S. electricity supply and, conversely, the importance of the electricity industry to the U.S. coal industry cannot be overstressed. In 1985, U.S. coal-fired power plants accounted for 84 percent of the nation's coal consumption. As shown in Table 14, coal-fired power plants accounted for 43.6 percent of the nation's installed capacity; oil-fired plants, 11.7 percent; gas-fired, 18.3 percent; water, 13.5 percent; and nuclear, 12.2 percent.

Coal-fired plants provide over 56 percent of the nation's electricity since many coal facilities are base load plants. It is expected that coal-fired power plants will continue to provide most of the nation's electricity in the future.

The shift to coal is largely explained by three factors. First, federal policy has favored the use of coal over alternative fossil fuels for use in power plants until very recently. The Energy Supply and Environmental Coordination Act of 1974 provided the Federal Energy Administration with the authority to require coal use in power plants, which led to the identification of plants

Table 14
Electrical Capacity and Generation in the United States, 1985

	Installed Capacity[a] (Percent)	Electricity Generated[b] (Percent)
Coal	43.6	55.5
Oil	11.7	4.9
Gas	18.3	12.3
Water	13.5	13.3
Nuclear	12.2	13.6
Other[c]	0.6	0.1

[a] U.S. Energy Information Administration, 1985. *Inventory of Power Plants in the United States*, Table 1.

[b] U.S. Energy Information Administration, 1985. *Electric Power Monthly*, Table 3.

[c] Includes geothermal, wood, waste, wind, and solar.

that should be converted from oil to coal. Another federal action, the 1978 Power Plant and Industrial Fuels Use Act, prohibited the use of oil and gas in new generating plants. Although exceptions were granted, the legislation was effective in reducing the number of oil and gas facilities completed, but it has since been terminated.

Second, there has been a loss of confidence in nuclear power. In 1974, the electric utility industry had plans for an additional 217,000 megawatts of nuclear capacity to be completed by the early 1990s. As of December 1982, 109,754 megawatts (100 units) of nuclear power capacity were canceled in the United States. The nuclear industry faced public opposition even before the 1979 partial meltdown at Three Mile Island; since then the Nuclear Regulatory Commission has enacted tougher plant safety standards. These new standards substantially raised the cost of completing nuclear plants under construction. Also, emergency evacuation planning requirements have led to the demise of the Shoreham nuclear plant in Long Island and problems with the Seabrook plant in New Hampshire.

Third, except in the past year, the delivered price of coal on a British thermal unit basis in most regions of the United States has become much less than oil and gas. According to several studies, the delivered price of coal relative to the delivered price of alternative fuels is the most important factor in determining the amount of coal consumed.

As discussed previously, coal-fired power plants account for 56.8 percent of the net electricity generation in the United States, and it is likely that coal's contribution will remain at about this level throughout the remainder of this century. The major advantage of coal over other fossil fuels is its great abundance and wide distribution in the United States, as illustrated in Fig 10. As of 1976, recoverable reserves exceeded 282 billion tons (Table 15). Coal production in the United States was a record 898 million tons in 1984 meaning that at the 1984 level of production the nation's coal reserves could last for more than 300 years. Furthermore, only New England and the Pacific Coast regions of the country lack substantial coal reserves.

Despite the availability of coal, it is not the most desirable fuel in many respects. Underground mining is dangerous and unhealthy to miners; surface mining disrupts the natural landscape. Because the energy content per ton is much lower for coal than for petroleum and natural gas, power plants require large quantities of coal; therefore, large quantities of coal must be transported and stored. Besides being expensive to handle and transport, the movement of coal often produces local particulate problems. The combustion of coal also creates environmental problems because of the release of gases and particulates. Finally, solid wastes (fly ash, bottom ash, and scrubber sludge) are produced by combustion and scrubbers. Disposal of wastes is an enormous problem in certain parts of the Northeast. Thus, at every stage of the coal resource production system, from mining through combustion, the use of coal is problematic.

The pulverized coal-fired power plant is the most common system for coal combustion in the United States for electric utilities. First used in the 1920s, pulverized-coal technology was refined throughout the twentieth century. Crushed and dried coal is blown into the combustion chamber where it burns at temperatures of about 2700° F. Heat energy from the flame heats the furnace walls by radiation to raise the temperature of water to create high-pressure steam. The steam rotates a turbine to drive a generator, thus generating electricity.

In the early part of the century, the efficiency of power plants was about 5 percent compared to close to 40 percent at plants completed in the late 1960s. This greater efficiency was accomplished by new burning and coal handling techniques as well as by larger units. In the 1970s, however, the efficiency of power plants declined somewhat because of the required environmental controls, which have also contributed to an increase in the cost of conventional power plants-almost 40 percent of the capital expenditures on a new plant are for environmental equipment. Electrostatic precipitators were first required to capture particulates from flue gases. As flue gases pass through a precipitator, particles are charged by passing between high-voltage electrodes and electrically grounded collecting plates where they collect and can be captured (Office of Technology Assessment 1979, 94). Electrostatic precipitators can capture 99.9 percent of the particulates. Capturing particulates is a minor problem,

Figure 10
Coalfields of the Conterminous United States

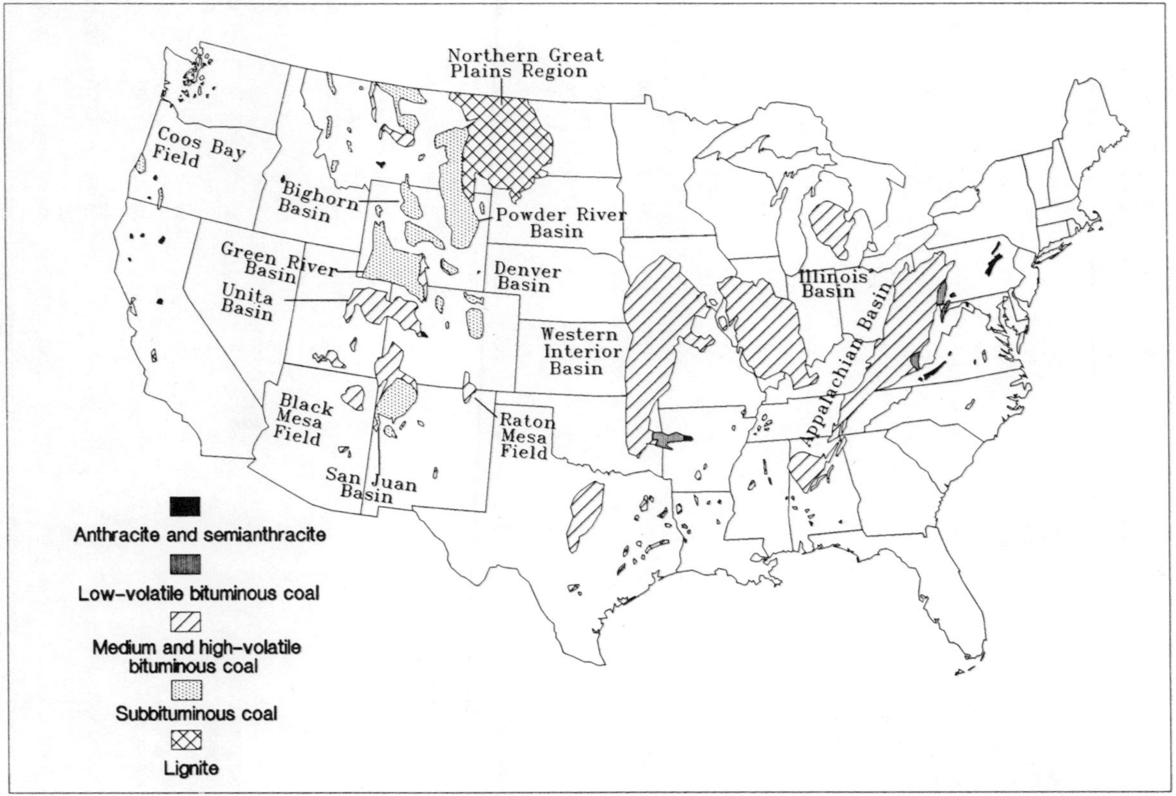

Source: Averitt, 1974.

Table 15
Recoverable Coal Reserves As of January 1, 1976

State	Heat Content (quadrillion Btu)			Recoverable Reserves (million tons)		
	Under-ground	Surface	Total	Under-ground	Surface	Total
Ohio	180	107	287	7,500	4,900	12,400
Pennsylvania	418	28	446	16,700	1,200	17,900
E. Kentucky	136	84	220	5,200	3,600	8,800
Virginia	53	17	70	2,000	700	2,700
W. Virginia	516	100	616	19,100	4,100	23,200
Maryland	14	3	17	500	100	600
Alabama	27	26	53	1,000	1,100	2,100
Tennessee	9	6	15	400	300	700
TOTAL APPALACHIA	1,353	372	1,725	52,400	16,000	68,400
Illinois	682	257	939	30,300	12,700	43,000
Indiana	117	29	146	5,100	1,400	6,500
W. Kentucky	118	69	187	4,800	3,200	8,000
TOTAL E. INTERIOR	917	355	1,272	40,200	17,300	57,500
Arkansas	4	3	7	100	100	200
Iowa	23	8	31	1,000	400	1,400
Kansas	--	19	19	--	800	800
Missouri	18	57	75	800	2,900	3,700
Oklahoma	18	8	26	700	300	1,000
TOTAL W. INTERIOR	62	95	157	2,600	4,500	7,100
Montana	898	794	1,692	40,400	39,700	80,100
N. Dakota	--	118	118	--	8,100	8,100
Wyoming	421	404	825	18,000	19,000	37,000
S. Dakota	--	4	4	--	300	300
Colorado	159	61	220	7,100	3,000	10,100
Utah	91	5	90	3,600	200	3,800
Arizona	--	5	5	--	300	300
N. Mexico	29	42	71	1,200	2,000	3,200
Texas	--	52	52	--	·2,500	2,500
Washington	--	--	--	600	400	1,000
Alaska	--	--	--	3,100	600	3,700
TOTAL WEST	1,598	1,485	3,083	74,000	76,100	150,100
TOTAL U.S.	3,931	2,307	6,238	169,000	113,900	282,900

Source: U.S. Office of Technology Assessment, 1979. The Direct Use of Coal, Washington, D.C.:U.S. Government Printing Office.

Options for Increasing Electricity Supplies

however, compared to reducing the sulfur dioxide emissions by using wet scrubbers. Wet scrubbers, for instance, use about 5 percent of the plant's output to operate pumps and fans and to reheat the flue gas. Also, scrubbers often experience operation problems which bring the plant down. Scrubbers also are very expensive, consume great quantities of water, and result in the generation of scrubber sludge, which presents a disposal problem in many areas.

All pulverized coal-fired power plants which began construction after 1978 had to have scrubbers, irrespective of the sulfur content of the coal consumed. Most existing coal-fired power plants do not have scrubbers. A utility considering the addition of a new coal-fired power plant would have to give serious thought to the construction of a pulverized-coal plant with a scrubber because there are not many alternatives that are ready at the commercial scale in the sizes required for most systems. There is great interest in other coal combustion technologies largely because of the unhappiness of many utility planners with scrubber technology. These technologies include fluidized-bed combustion and IGCC.

Fluidized-bed combustion involves the suspension of finely crushed coal with crushed limestone in such a manner that the mixture behaves like a fluid. The mixture is heated causing the coal to burn, and the resulting sulfur dioxide reacts with the limestone (or dolomite) to produce dry calcium sulfate (Office of Technology Assessment 1979, 102). In addition to capturing the sulfur oxides, the fluidized-bed method produces less nitrogen oxides because of the lower bed temperatures used (1,500° to 1,800° F) compared to conventional boilers (2,700° to 3,000° F) and less oxygen enrichment. Since the fluidized bed burns at a temperature below that which would cause coal ash to melt, the plants are capable of burning a wider variety, and often lower quality, of coal than conventional boilers which experience fouling of heat exchange surfaces from the melting of ash. An important drawback of the fluidized bed is that bed temperatures must be held relatively constant, making it difficult to maintain operations during low-load periods. Two major types of fluidized-bed units are being developed: atmospheric and pressurized.

In 1976, the Monongahela Power Company began experiments with a 30-megawatts atmospheric fluidized-bed plant in Rivesville, West Virginia. A 160-megawatts plant has been built in Kentucky sponsored by the Tennessee Valley Authority, Duke Power, and the Commonwealth of Kentucky. Other smaller plants are in operation. The technology seems to be best suited for use in small power plants. Utilities using the fluidized-bed will probably need to add multiple units of about 100 megawatts in order to bring on-line substantial blocks of generating capacity.

A pressurized fluidized bed is different in that it maintains the combustion bed at between 4 and 16 atmospheres of pressure (Office of Technology Assessment 1979, 103). This increased operating pressure allows the furnace to be smaller, increases combustion efficiency, and improves the rate at which sulfur oxides are captured so that less limestone (or dolomite) is required. There is interest in developing a pressurized fluidized bed with combined cycle operation using gas turbines. This would further increase plant efficiency, but the problem of removing corrosive materials from the flue gas to avoid turbine corrosion must be solved (Office of Technology Assessment 1979, 104).

The IGCC technology, a clean coal technology, has received widespread attention because of a successful demonstration plant. Southern California Edison Company's 100 megawatts Cool Water plant at Daggett, California, was completed in 1984, after a two-and-one-half year construction period at a cost of $294 million (Spencer, Alpert, and Gilman 1986, 233). The process involves crushing coal and slurrying and gasifying the mixture. Slag is removed from the gasifier releasing radiant heat to raise steam. The gas is desulfurized to remove from between 95 and 99 percent of the sulfur. Once sulfur and other impurities have been removed, the gas is heated and burned in a combustion turbine.

The plant has been successful in burning coal cleanly, and its emissions are similar to those of gas-fired plants. Test burns with Utah, Illinois, and West Virginia coals have been conducted. The plant can also be constructed in stages that provide utilities with a means to add capacity as demand grows. Gas turbines could first be built at a site using natural gas. As electricity demand grows, a coal-gasification facility could be added. A drawback to the technology is its heat rate. Initial operations recorded a heat rate of 11,300 Btu/kWh; it is hoped that the heat rate can be lowered to 9,000 Btu/kWh to meet conventional power plant standards (Spencer, Alpert and Gilman 1986, 610-11). A 63-megawatts Appalachian Project, which was planned for Somerset County, Pennsylvania was projected to have a 9,000-Btu/kWh heat rate (Buckman 1988).

There are also other ways to produce electricity from coal. Solvent-refined coal can also capture sulfur oxides. In this system, crushed coal is dissolved in a solvent at a moderately high temperature and pressure, and the mixture is treated with hydrogen to remove sulfur. The solution is filtered to remove ash minerals, and the solvent is removed to recover the clean product (Office of Technology Assessment 1979, 101). Two small pilot plants have been operating, one in Alabama and another in Washington. Commercial sized plants have been proposed.

Other technologies that do not appear eminent are magnetohydrodynamics (MHD) and fuel cells. MHD involves the flow of hot gas through a magnetic field at high velocity. An electrical current can be produced because the high-temperature gas conducts electricity, which can be captured by electrodes. When this is done using coal, however, ash slag accumulates in the system reducing the electrical current output. Fuel cells are also unlikely to make an important contribution to the nation's electricity supply in the near future. In a fuel cell, two electrodes are immersed in an electrolyte. Fuel cells are used to power many space ventures. Clean, low-Btu gas from coal could be used.

Nuclear Power

It was once believed that, by the early twentieth century, nuclear power would provide most of the nation's electrical generating capacity. It was thought that fewer coal-fired plants would be ordered because nuclear power plants would be a cheaper and cleaner source of electricity. Today, however, the future for nuclear power is very bleak and is not really an important alternative for those interested in adding new base load generation. The last nuclear power plant was ordered in the United States in 1978; the last one that will be completed was ordered in 1974. Because it has taken so long for many of these plants to be completed, there has been a recent surge in new nuclear-powered capacity coming on-line in the United States, which is adding to already high reserve margins. In the long term, as some existing nuclear units are retired, nuclear capacity will only decrease and decline in importance as a source of electricity. Some nuclear-power advocates are calling for "a second nuclear era," but even these individuals recognize that new reactor designs, and probably smaller units, will be necessary if nuclear power is to be publicly acceptable and economically competitive (Spinrad 1988, 707). In any case, a revival of the industry is not expected until the next century.

Nuclear power plants account for about 12 percent of the nation's total generating capacity. Over the period from 1986 to 1995, almost 43 percent of the new generating capacity to be brought on-line in the United States is scheduled to be nuclear (compared to 45 percent for coal). There are now ninety-eight operating nuclear power plant units in the United States. As shown in Table 16, there will be a decline in nuclear power plant construction. There are currently twenty-nine nuclear power plants under construction representing 32 gigawatts of generating capacity. Almost all of these units are scheduled to be completed by 1988. If all of these units are completed, there will be 127 nuclear units providing 112 gigawatts of generating capacity when the nuclear construction activity ends. This is well below previous forecasts of the importance of nuclear power. In 1974, there were orders placed for over 217 gigawatts of nuclear capacity, which was to have been added to the existing 30 gigawatts of capacity (U.S. Energy Information Administration 1983, 5).

Electric utilities began to cancel orders for nuclear power plants in 1972. Since that time, over 100 nuclear units have been canceled, far exceeding the number of coal-fired units canceled. In a study conducted by the U.S. Energy Information Administration, several reasons accounting for these plant cancellations were identified. The primary explanation was that the forecasted growth in peak load was revised downward. This factor would affect non-nuclear units as well as nuclear units, but nuclear units were more vulnerable because it was taking so long for them to be completed and placed into operation. In 1972 and 1973, it took, on an average, six years to construct and place into commercial operation a new nuclear power plant. By 1984 and 1985, it took over eleven years

Table 16
Nuclear Power Reactors Currently Under Construction by Expected Year of Entry Into Operation

Expected Year of Entry into Commercial Operation	Number of Units	Total Summer Capability (Gwh)
1986	14	15.0
1987	9	10.1
1988	2	2.3
1989	1	1.2
1990	1	1.1
1994	1	1.2
1996	1	1.2
Total	29	32.1

Source: U.S. Energy Information Administration. 1986, Nuclear Power Plant Construction Activity 1985.

for a nuclear plant to be constructed and placed on-line. It now takes almost thirteen years for a nuclear power plant to be constructed and placed into operation (U.S. Energy Information Administration 1986, 14). This problem accounts for another major explanation of nuclear power plant cancellation; increasing financial problems.

Because of the extended lead times, combined with new design requirements, the cost of constructing nuclear power plants increased alarmingly. The average construction cost per kilowatt increased from $161 (nominal dollars) in the 1968-71 period to $1,542 in the 1979-84 period (U.S. Energy Information Administration 1986, 13). The seven units that entered commercial operation in 1985 averaged $2,466 per kilowatt, and the units going on-line in 1986 are estimated to cost $3,230 per kilowatt electric. It is believed that about 75 percent of the increase in construction costs are attributed to increases in the quantities of land, labor, material, and equipment used to construct nuclear power plants. The remaining increases in construction costs are related to increases in real financing charges; escalation in the prices of land, labor, material, and equipment; and elongation of the construction lead times (U.S. Energy Information Administration 1986).

The dim prospects for nuclear power are echoed by the investment community, which was particularly nervous about nuclear power after the Three Mile Island accident in 1979. Investors rate nuclear utilities as riskier investments than non-nuclear utilities, according to a study conducted by the U.S. Energy Information

Administration. Investment houses such as Merrill Lynch and Salomon Brothers expressed concern to investors considering utilities with substantial nuclear commitments. The concern is traced to three factors: (1) the failure of many nuclear power plants to operate at anticipated utilization rates, (2) nuclear safety concerns, and (3) the cost of abandoned plants (U.S. Energy Information Administration 1984). In some cases, utilities were forced to purchase power from other systems because their nuclear plants were not operating at a high enough operating factor. Part of the problem results from extended outages and inspections which halt power production. The cost of the Three Mile Island cleanup, estimated at $1.5 billion, required utilities to increase their coverage in case of accident. It is now considered prudent to hold $1 billion in insurance in case of an accident. The private insurance pool, however, will provide only up to $500 million in insurance; therefore, nuclear utilities have formed a collective self-insurance company known as Nuclear Electric Insurance Limited. This provides additional exposure for nuclear utilities because these utilities would be assessed a premium to cover an accident. The Price Anderson Act, which limited liability from a nuclear accident to $590 million, is now being extended to provide greater coverage. Finally, the cost of project abandonment adds risk to utilities who are constructing nuclear power plants. The largest default of municipal bonds in history occurred when the Washington Public Power Supply System (WPPSS) canceled unfinished plants, resulting in over $2 billion in abandonment costs. In 1968, WPPSS began to plan for forty coal plants and twenty nuclear plants. By 1982, it was realized that the cost of finishing five nuclear power plants would be $24 billion, and that it was doubtful if the electricity was actually needed (Rudolph and Ridley 1986, 3). The cost of plant decommissioning is also expected to be substantial. In 1987, Pacific Gas and Electric Company was ordered by the California Public Utilities Commission to begin collecting $26 million annually for decommissioning its two Diablo Canyon units (Bernow and Biewald 1987, 14). However, the actual costs of decommissioning will be unknown until the industry begins to decommission large plants.

Other reasons accounting for nuclear power plant cancellations, and the undesirability of investing in nuclear power plants, include regulatory uncertainty, the lack of economic advantage with respect to coal-fired facilities, and denial of certification by state commissions. Following the Three Mile Island accident, the Nuclear Regulatory Commission promulgated new regulations for the design and operation of nuclear plants. Perhaps the most demanding change is the completion and approval of emergency evacuation plans prior to licensing. This has proven to be a major stumbling block for certain reactors in populated areas, particularly Shoreham in Long Island and Seabrook in southern New Hampshire.

The emergency evacuation procedures requirements provided local governments and citizens with more control over nuclear power licensing. Before a plant can be licensed, the Federal Emergency Management Agency (FEMA) and the

Nuclear Regulatory Commission (NRC) must approve an evacuation plan. Local officials must be involved in emergency planning in order for the plan to be approved. Licensing can be deterred if local officials refuse to approve or participate in emergency planning procedures. This was the first time that local governments could exercise control over nearby nuclear power plants (Axelrod and Wilson 1986, 316). In February 1983, the Suffolk County legislature passed a resolution supporting the county executive's position that the area around the Shoreham nuclear power plant could not be safely evacuated, given traffic patterns and topography, and that no evacuation plan could be prepared (Axelrod and Wilson 1986, 324). Even though the governor could have overridden this resolution, Governor Cuomo did not impose a state plan. As it turned out, the plant will be turned over to the State of New York and will not go into operation, and Long Island Lighting Company will be able to raise rates and obtain tax breaks. Thus, a local community was able to thwart a nuclear plant from going on-line.

Because of these many problems, it is doubtful if any utility will order a new nuclear power plant in the near future. Plants coming on-line now will increase nuclear power's contribution temporarily. Also, utilities will tend to operate these plants continuously to capture the high capital costs in their construction. Nuclear power cannot be considered at this time to be a realistic alternative to additional base load coal-fired capacity for providing additional generating capacity.

RETROFITTING EXISTING CAPACITY

The nominal life of most conventional power plants is from thirty to forty years. According to the Atomic Energy Act of 1954, a nuclear power plant's operating license would expire forty years after the construction permit was issued. Because it was taking so many years to complete nuclear power plants, in 1982 it was decided to begin the forty-year period after the operating license was issued. Since current nuclear licenses, representing 50 gigawatts of nuclear power, will expire before 2008, there is discussion of extending the life of nuclear plants, even though there is no formal list of requirements needed to obtain extensions to the operating life of a nuclear power plant (Williams 1986). Aging coal plants are not so strictly regulated, but they do experience declining performance. More outages occur and the likelihood of accidents, such as lethal steam line failures, increase, making it necessary to invest in the plant or retire it. The average age of U.S. generating capacity has been increasing because there has been so little generating capacity added over the past ten years. By 1995, over 30 percent of the nation's coal-fired generating capacity will be over thirty years old (U.S. Energy Information Administration 1985).

Many utilities are now considering extending the life of their larger facilities instead of retiring them and building new plants. There is very little capacity retirements scheduled to 1995. Of the total U.S. generating capacity of 698,099 megawatts, only 13,499 megawatts are scheduled for retirement by 1995, or less than 2 percent of the total capacity. Table 17 indicates that gas- and petroleum-fired facilities account for 83 percent of the planned retirements. Only 1,299 megawatts of coal-fired capacity is to be retired by 1995.

In the past, power plants that were retired were replaced by larger facilities which enjoyed higher efficiencies through the use of new power plant technology. However, modern power plants are not too much different than those that were completed in the early 1960s, and they may even have lower efficiencies because of flue gas desulfurization equipment. Also, the average capacity of power plant units increased throughout the century and levelled off at about 500 megawatts. Power companies discovered that few economies were realized by constructing facilities greater in size than this. As the 25-megawatts plants built in the 1930s reached the end of their useful operating life, the power company could replace the plants with much larger, more efficient units. It makes much less sense to retire a 300 megawatts plant built in the 1950s with a new plant that is not much larger and employs the same basic technology.

There are several other reasons why refurbishing an existing plant is much more attractive today. The primary reason is that refurbishing an existing plant is much cheaper than building new generating capacity. Life extension costs range from $200 to $300 per kilowatt compared to a price range of from $1,200 to $2,000 per kilowatt for installing new capacity (Smock 1987, 18). Refurbishing an existing facility can be accomplished much more quickly than building a new power plant. It may take over eight years to license and construct a new power plant; refurbishing can be completed in less than a year. Siting a new plant also requires permits and public hearings that are avoided by activities at existing power plants.

Also, older coal-fired power plants grandfathered from the 1979 Clean Air Act Amendments, may be retrofitted without having to install a scrubber. New capacity must use a flue gas desulfurization system. The extended operating life of many large coal-fired power plants is one reason why acid rain legislation may target the older plants that are grandfathered.

Many utilities now have vigorous life extension programs. Pennsylvania Power and Light, for instance, has budgeted $1.25 million annually for such activities. The utility is planning to replace turbines, upgrade boilers, replace control system components and complete some generator rewinds (Smock 1987, 21).

Life extension makes it more difficult to forecast the need for new generating capacity. Not all facilities are candidates for life extension work, and refurbishing existing power plants does not add new capacity to the system. Life extension may delay the day when new generating capacity is needed, but it does not avert the need for the provision of new capacity.

Table 17
Planned Capacity Retirements by Year As of December 1985 (nameplate capacity, MW)*

Year	Coal	Petro	Gas	Nuclear	Hydro	Other	Total
1986	41	631	144	--	1	10	827
1987	110	134	574	--	8	--	826
1988	33	387	538	--	5	--	963
1989	190	145	279	--	8	--	622
1990	482	546	854	--	1	110	1,993
1991	198	665	934	--	--	--	1,797
1992	--	622	1,295	--	--	--	1,917
1993	100	563	887	--	--	--	1,550
1994	144	101	989	860	--	--	2,094
1995	2	121	786	--	--	--	909
TOTAL	1,300	3,915	7,280	860	23	120	13,498

*Tables may differ when summed because of rounding.

Source: U.S. Energy Information Administration, 1986, Inventory of Power Plants in the United States, 245-46.

COGENERATION

Cogeneration involves the production of electricity and heat from the same energy source. There are three types of cogeneration: (1) district heating, (2) residential, commercial, institutional cogeneration, and (3) industrial cogeneration. Over 60 percent of the energy used in the generation of electricity is lost to the environment. District heating systems capture some of this heat and pipe it to nearby consumers who use it for heating and hot water purposes. Only fifty-nine district heating systems are operating in the United States; all but one provides steam (Hu 1985, 269). District heating was once more popular. In 1909, 150 U.S. utilities had district heating systems (Hu 1985, 269). Of the remaining systems, about forty have been in existence since the turn of the century (Office of Technology Assessment 1983, 164). District heating is much more widely used in northern European countries than it is in the United States. In the United States, district heating systems are declining in popularity, and few utilities favorably consider the alternative because of the

Options for Increasing Electricity Supplies

expense and the difficulty and disruptiveness of installing pipes to serve dispersed customers with low-temperature steam.

There are few examples of residential and commercial cogeneration systems operating in the United States, but many institutions do produce both power and steam. Universities, prisons, and military bases often have a central power plant which generates power and distributes steam to the institution's buildings via underground pipes.

Industrial cogeneration has seen the greatest expansion and offers the largest potential contribution to the nation's installed generating capacity. Many industrial processes require heat, and cogeneration involves adding a turbine-generator system to a boiler to produce electricity to be used on site or for sale back to the local electric utility. Industrial cogeneration was once much more common than it is today. In 1900, almost 60 percent of total United States electric generating capacity was located at industrial sites, although all of these were not cogenerating facilities (Office of Technology Assessment 1983, 3). Throughout the twentieth century, cogeneration declined as electric utilities expanded their transmission and distribution network and as economies of scale were realized by the construction of larger utility power plants. It became cheaper and less troublesome for industry to purchase power from an electric utility. By 1980, industrial cogeneration accounted for only 3 percent of the nation's generating capacity (Office of Technology Assessment 1983, 3).

As of 1982, there was an estimated installed capacity of 14,858 megawatts of industrial cogeneration. As shown in Table 18, five industries accounted for 87 percent of the industrial cogeneration: (1) pulp and paper, (2) primary metals, (3) chemicals, (4) petroleum refining, and (5) food. These industries consume a large amount of thermal energy in their production processes.

According to the U.S. Office of Technology Assessment, there exists great technical potential for industrial cogeneration but little market potential because of economic and institutional considerations (Office of Technology Assessment 1983, 32). For many industries, cogeneration cannot compete with the price of electricity supplied by the electric utility. Also, many industries are reducing their energy use by conservation and industrial changes that reduce the waste heat generated. In the steel industry, for example, the replacement of open-hearth furnaces by electric-arc processes reduces the cogeneration potential (Office of Technology Assessment 1983, 33).

As electricity prices rose in the 1970s and there was a national effort to promote alternatives to the construction of new energy sources, cogeneration realized renewed growth. The Public Utilities Regulatory Policies Act of 1978 (PURPA, 95-617) encouraged cogeneration because producing heat and electricity at a cogeneration facility uses less energy than producing electricity and heat separately at two distinct facilities. Section 210 of PURPA, promulgated by Congress in 1980, and administered by the Federal Energy Regulatory Commission (FERC), requires

Table 18
Installed Industrial Cogeneration Capacity

Industry	Capacity (MW)
Pulp and Paper	4,246
Primary Metals	3,589
Chemicals	3,438
Petroleum Refining	1,244
Food	398
Other	1,943
	14,858

Source: General Energy Associates, *Industrial Cogeneration Potential Targeting of Opportunities at the Plant Site*, 1982. Cited in Office of Technology Assessment, 1983.

utilities to purchase power provided by qualified cogenerators at the utility's avoided cost, or another mutually agreed-upon price. The avoided cost, calculated by using the utility's capacity and energy cost, is that which would be incurred if the utility had to generate the power itself. Determining this cost has been problematic in many cases and often involves a charge to the utility for administrative expenses for accounting for the cogenerators' contribution (Stewart 1986, 25). The Pennsylvania Public Utility Commission, for example, ordered cogeneration developers in the Philadelphia Electric Company's territory to pay a $3.35 per kilowatt per month standby backup service charge in addition to energy and capacity charges (Devlin 1985, 46). PURPA also requires that the utility provide the cogenerator with backup power, when needed, at nondiscriminatory rates. In order to qualify under PURPA 210, the cogenerator must not be owned by more than 50 percent of equity by an electric company (Butler 1984, 85). The facility must meet certain operating and efficiency standards. The FERC requires cogeneration facilities that produce electric power before thermal energy (topping cycle cogenerators) to use at least 5 percent of its total energy output as useful thermal energy. Topping cycle plants must also meet standards that ensure that they are more efficient than facilities that produce electricity and heat separately (Butler 1984, 13; Office of Technology Assessment 1983, 77). Bottoming cycle plants which produce electricity from heat that would have been wasted have less stringent efficiency standards. If a cogenerator is a qualified facility under PURPA 210, then the firm can sell electricity without being classified as

Options for Increasing Electricity Supplies

an electric utility and being regulated by the FERC and the state public utility commissions.

Cogeneration is not likely to solve the nation's electricity supply problems, but it could be very important in providing short-term capacity in selected areas. According to a Dun and Bradstreet-TRW study, cited by the U.S. Office of Technology Assessment (1983), there exists in the United States a potential 39,344 megawatts of cogeneration capacity. The largest contributions can be added in areas with existing petrochemical industries, such as Texas and Louisiana. The report indicates that Texas could develop over 5,000 megawatts of cogeneration capacity. The west/south/central census region has 30 percent of the cogeneration capacity.

The availability of cogeneration also depends upon the avoided cost. Areas with high electricity prices will have more cogeneration capacity brought forth than areas with low electricity prices. New England and the Middle Atlantic states have had a large number of cogeneration projects largely because of the high avoided cost figure which many cogenerators can beat. These areas are dominated by oil-fired capacity; if the price of oil rises, cogeneration is likely to be an attractive alternative. However, no one really knows how much additional capacity can be brought forth at certain price levels. Cogeneration plants can generally be completed in a much shorter period of time than a new conventional coal or nuclear capacity. Vulcan Chemical Corporation, for example, completed a 110 megawatts cogeneration plant in Wichita, Kansas, in eighteen months (Smith and O'Donnell 1987, 41). Capital costs for cogeneration are also less than conventional alternatives, averaging between $800 and $1,300 per kilowatt of electric capacity (Zimmer and Jones 1986, 24).

The North American Electric Reliability Council (NERC) undertook a study of expected nonutility generator additions in North America and found that approximately 24,707 megawatts of new and existing facilities will be available by 1996 (North American Electric Reliability Council 1987, 20). In the Mid-Atlantic Area Reliability Council (MAAC) region, 3,312 megawatts are projected, representing 6.2 percent of the total capacity resources; 3,380 megawatts are projected for the Northeast Power Coordinating Council (NPCC) region, or 2.6 percent of the total capacity resources.

In coal-mining areas, there is interest in recovering "gob" from abandoned coal mines in cogeneration plants. In addition to providing a source of power, such developments reduce an environmental problem. There is currently much activity of this type in Pennsylvania, but some plans are not going forward because of the lack of electricity transmission lines to move power from isolated coal-mining areas.

Utility representatives are concerned that nonutility generators could pose system reliability problems and complicate the planning process for new utility capacity. Most cogeneration facilities provide less reactive power than is needed to maintain network voltage levels which forces the utility to operate their own units (North

American Electric Reliability Council (NERC) 1987, 29). For the most part, reactive power must be provided near the load center. Utilities are concerned that the location of cogeneration facilities could make transmission of their power difficult or burden systems scheduling. If cogeneration plants are clustered, the utility may lack transmission reinforcements whereas highly dispersed, small units make system control and scheduling difficult. Also, because most of these units are new, utility planners lack the data needed to determine their reliability and availability. Likewise, cogenerators may be phased out quickly without the utility's having an opportunity to arrange substitute power.

CONCLUSION

Very few conventional power plants have been ordered in recent years. There are new technologies that offer options for the use of coal for electricity generation, but it is unlikely that utilities will invest in new nuclear generating facilities in the near future. Utility planners have more options available to them than they did in the past and are operating in a new regulatory environment as public utility commissions deal with cogeneration facilities. The life extension of existing facilities and the development of cogeneration facilities will postpone capacity expansion decisions, but efforts must be taken soon in New England and the Mid-Atlantic states to obtain new capacity. In the next chapter we discuss power trade as a means to provide additional capacity in these areas.

BIBLIOGRAPHY

Axelrod, Reginia S., and Hugh A. Wilson. 1986. "Citizen Participation and Nuclear Power: The Shoreham Experience." The Journal of Energy and Development. 11, 2:311-31.
Bernow, Stephen S., and Bruce E. Biewald. 1987. Nuclear Power Plant Decommissioning: Cost Estimation for Planning and Rate Making." Public Utilities Fortnightly. 120, 9:14-20.
Buckman, J. W. May 10, 1988. "The Appalachian Project." Paper presented at the Eighth Annual Gasification and Gas Stream Cleanup Systems Meeting, Morgantown, W. Va.
Butler, Charles H. 1984. Cogeneration. New York: McGraw-Hill Book Company.
Devlin, J. 1985. "Cogeneration Development in Pennsylvania." Alternative Sources of Energy. 85: 46-47.
Haman-Guild, Renee, and Jerry L. Pfeffer. 1987. "Competitive Bidding for New Electric Power Supplies: Deregulation or Reregulation?" Public Utilities Fortnightly. 120, 6:9-20.
Hu, S. David. 1985. Cogeneration. Reston, Va.: Reston Publishing Company.

Meade, William R. 1987. "Competitive Bidding and the Regulatory Balancing Act." *Public Utilities Fortnightly*. 120, 6:22-30.

North American Electric Reliability Council. 1987. *The Future of Bulk Electric System Reliability in North America 1987-1996*. N.J.:(NERC).

Office of Technology Assessment. 1983. *Industrial and Commercial Cogeneration*. Washington, D.C.:U.S. Government Printing Office.

---. 1979. *Direct Use of Coal*. Washington, D.C.:U.S. Office of Technology Assessment.

Rudolph, Richard, and Scott Ridley. 1986. *Power Struggle: The Hundred Year War Over Electricity*. New York:Harper and Row.

Smith, C. Kent, and John O'Donnell. 1987. "110-MW Cogeneration Plant Designed and Built in 18 Months." *Power Engineering* 91:41-43.

Smock, Robert. 1987. "Power Plant Owners 'Phase in' Life Extension." *Power Engineering*. 91:18-23.

Smock, Robert, and Mona Reynolds. 1987. "New Utility Generating Plants: When, Where and by Whom." *Power Engineering*. 91:15-21.

Spenser, D. F., Alpert, S. B., and H. H. Gilman. 1986. "Cool Water: Demonstration of a Clean and Efficient New Coal Technology." *Science*. 232:609-12.

Spinrad, Bernard I. 1988. "U.S. Nuclear Power in the Next Twenty Years." *Science*. 239, 12:707-8.

Stewart, Robert D., Jr. 1986. "The Law of Cogeneration in Oklahoma." *Public Utilities Fortnightly*. 118, 11:22-26.

U.S. Energy Information Administration. 1986. *Nuclear Power Plant Construction Activity 1985*. Washington, D.C.:U.S. Department of Energy, U.S. Energy Information Administration (DOE/EIA).

---. 1985. *Inventory of Power Plants in the United States*. Washington, D.C.:U.S. Department of Energy, U.S. Energy Information Administration (DOE/EIA).

---. 1984. *Investor Perceptions of Nuclear Power*. Washington, D. C.: U.S. Department of Energy, U.S. Energy Information Administration (DOE/EIA).

---. 1983. *Nuclear Plant Cancellations: Causes, Costs and Consequences*. Washington, D.C.:U.S. Department of Energy, U.S. Energy Information Administration (DOE/EIA).

Williams, Lawrence. 1986. "Planning Needed for Nuclear Plant Life Extension." *Public Utilities Fortnightly*. 118, 5:24-28.

Zimmer, Michael J., and Beverly E. Jones. 1986. "Cogeneration: Boon or Bane to Consumers?" *Public Utilities Fortnightly*. 117, 12:23-29.

5

The Power Trade Alternative

Frank J. Calzonetti and Muhammad A. Choudhry

As the electric industry grew throughout this century, neighboring utilities interconnected to provide mutual economic and reliability benefits. There are now three major interconnected systems in the United States: the eastern interconnection; the Texas interconnection; and the western interconnection. These are shown in Fig 11.

The interconnected transmission system in North America is organized by control areas which match an area's generation to the load and monitor the area's power imports and exports (Federal Energy Regulatory Commission 1981, 26). Individual utilities within these control areas also control its generation and power trade, but the country is divided into three systems, each of which has a number of energy control centers. The eastern interconnected system, which extends from the Canadian Atlantic provinces, south to Florida and west to the Great Plains, has about 100 control centers. The western interconnected system has about thirty-four control centers, and the Texas interconnected system has about ten (Federal Energy Regulatory Commission 1981, 26). Reliability is the primary concern in the operation of these interconnected systems, and reliability is not sacrificed to achieve greater economies. It must be understood that an event in one location can affect the entire interconnected system. A failure in Pennsylvania could conceivably result in a cascading failure that could knock out power from the Atlantic to the Rocky Mountains.

The purpose of electricity transmission lines is to deliver electric energy from power plants to load centers in a reliable and cost effective manner. As power plant sizes increased throughout the 1950s, 1960s, and 1970s, more power was being transmitted over longer distances in the United States and at higher voltages.

Voltages in the 100-to-150-kilovolt range were used as early as the 1920s. In 1952, American Electric Power (AEP) installed a 345-kilovolt line. A 735-kilovolt line was energized in 1965 by Hydro-Quebec to transfer power from

Figure 11
Interconnections of the North American Electric Reliability Council

remote hydroelectric plants to load centers in the South of Canada. The first 765-kilovolt line was built by AEP and was activated in the 1970s. There are plans to construct 1100-kilovolt and 1500-kilovolt lines, but these have been postponed. Today, with more than 167,000 miles of transmission lines at 270-kilovolts or greater, the existing transmission network in North America is by far the most complex and sophisticated system in the world.

Fig 12 is a map of extra-high-voltage (EHV) transmission lines (above 345-kilovolts) in the contiguous United States and Canada. Most of the lines involve alternating current (AC) transmission: There are 179,161 circuit miles of AC transmission and 3,527 circuit miles of direct current (DC) transmission, notably in Canada and in the Far West; these lines are tied to the interconnected system by converter stations. In the interconnected AC transmission network, power moves instantaneously from the point of generation to the point of consumption across all available transmission paths. Control centers cannot direct power along a particular path, as is the case with DC transmission.

The power transmitted at a given power factor is proportional to the product EI where E is line-to-line voltage and I is the line current. Thus, for a given power to be transmitted, the higher the voltage used, the lower will be the power losses ΔP_L:

$$\Delta P_L = 3\ RI^2 \tag{1}$$

where: $$R = \frac{L}{\sigma A} \tag{2}$$

σ = conductivity of conductor (mile/$\Omega \cdot$ mm^2)
A = cross-sectional area (mm^2)
L = length of the line (miles).

The total power transmitted by a three-phase line is

$$P = (3)^{1/2}\ EI\ \text{Cos}\phi \tag{3}$$

where cosϕ is the power factor.

Substituting for I from equation (3) in equation (1) results in

$$\Delta P_L = \frac{L}{\sigma A} * \frac{P^2}{E^2\ (\text{Cos}\phi)^2}. \tag{4}$$

The above equations show that power losses ΔP_L are related to L, A, σ, P, E, and Cosϕ (power factor). The factors having a significant effect on power losses are conductor size A and voltage level E. One can thus determine an economic balance with respect to power losses, conductor size, and voltage level. Some other technical parameters which are to be considered in designing a transmission line are stability, voltage profile, current limitation, and radio interference.

Figure 12
North American Extra High Voltage Transmission Lines

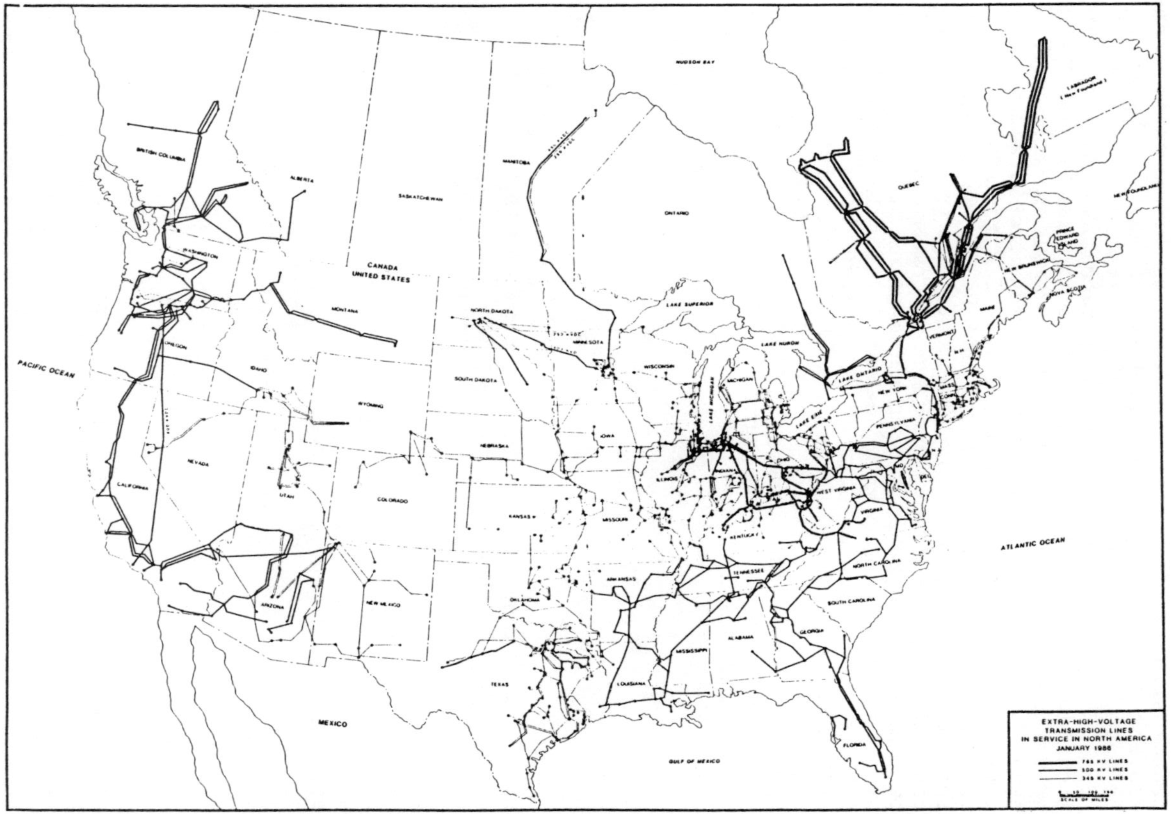

Stability

An AC system is stable if it can operate with all the synchronous generators connected to a remote DC system through a transmission line. The power transferred through the line is given by

$$P = \frac{E_g E_s}{X_L} \sin\delta \qquad (5)$$

where E_g and E_s are voltage magnitudes of the sending end and the receiving end, X_L is the reactance of the transmission line and other components, and δ is the difference of voltage angles of two buses. The power transferred through the line is directly proportional to the $\sin\delta$ and inversely proportional to X_L. The maximum power occurs when $\delta = 90°$. However, δ is kept at 30° to 45° due to stability considerations. The reactance X_L increases with the length of the line and limits the power transferred through long AC transmission lines (see Fig 13).

Voltage Considerations

Two of the more serious problems in EHV transmission are the production and consumption of reactive power. EHV transmission lines consume reactive power when they are carrying more than a natural load and produce reactive power when their load is less than a natural load. Natural load of a line is related to the voltage, inductance, and capacitance of the line. The voltage variation of the line terminals is kept to within 5 percent by using a reactive power generation source (capacitor band) and a reactive power consumption source (inductor). The voltage of the receiving end increases during no load or light load and decreases during heavy load conditions.

Another problem in EHV lines is the switching surge voltage. The peak value of the switching surge voltage may reach two or three times the normal crest voltage. Therefore, insulators and other components on the line should be able to withstand this voltage.

Current Limitation

As the resistance power loss ($\Delta P_L = RI^2$) increases the conductor temperature, the current through the conductor should be kept below the maximum permissible value. In most of the EHV transmission lines, stability limits are reached before the thermal loading limit.

Improvements to the Existing Grid

The power flow through an AC transmission line is inversely proportional to its reactance. The reactance of the line can be reduced by inserting capacitors in series

**Figure 13
Circuit Model of Two Machines Used for
Stability Consideration**

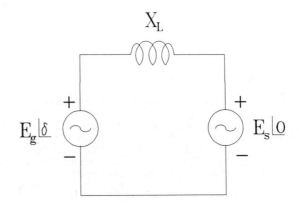

E_g = voltage magnitude of the sending end
E_s = voltage magnitude of the receiving end
X_L = the total inductive reactance between E_g and E_s
δ = angle of the sending end voltage

with the line (Dunlop, Gutman, and Marchenko 1979, 606). The series compensation allows the transfer of more power over long AC transmission lines. The voltage variation in transmission lines can be controlled using shunt reactors, shunt capacitors, and thyristor-controlled static volt-ampere reactive compensators (Hingorani 1986). Short transmission lines may reach their thermal loading limit before the stability limit. Power through a transmission line can be increased during cool weather if the transformers and other equipment at the substation do not reach their thermal limit. On-line monitoring of transformer-winding hot-spot temperature may allow more transfer of power through the transmission line.

The power through the AC transmission line is proportional to the difference of phase angles between the sending end and the receiving end. A phase-shifting transformer changes the flow of power through the line by changing the phase angle. The line current capacity of a transmission line can be increased by suspending additional conductors along with the existing one. This may require strengthening the structure and foundation to support the additional weight. Another approach to increase the transfer of power through a line is to upgrade the voltage of the existing low-voltage lines. This may require obtaining wider rights-of-way, installing better insulators, strengthening the foundation, and raising the transmission line to meet electric and magnetic field specifications.

HIGH-VOLTAGE DC TRANSMISSION

High-voltage DC (HVDC) transmission has been used to transmit power under water (there is a proposal to connect New England to Nova Scotia by an underwater cable), to transmit power over great distances on land, and to make asynchronous connection between two systems. There has been an increased interest in DC transmission throughout the world since the 1954 successful operation of the Gotland DC link to the Swedish mainland. This link transmits 20-megawatts at 100-kilovolts through the sea and earth. A full-scale 720-megawatts, ±400-kilovolts, 294-mile DC transmission line between a hydroelectric plant at Volgograd and Donets Basin in the Soviet Union was energized in 1962. A DC link connects the electrical systems of France and England. This link transmits 160-megawatts of power through two single-conductor submarine cables at ±100-kilovolts. The power can flow in either direction and can be set at the desired value.

Fig 14 shows the location of North American DC schemes which are operational or are under active consideration (Reeve 1984, 43). Table 19 shows the capacity and voltage rating of these schemes. All the HVDC projects before 1972 used mercury arc valves to connect AC power into DC power (rectifier) and to convert DC power into AC power (inverter). Solid-state valves were first used in 1972 at the Eel River project. These valves have better performance than mercury arc valves.

The DC links used in various projects are monopolar, bipolar, or homopolar (Kimbark 1971, 11). The monopolar

Figure 14
North American DC Schemes

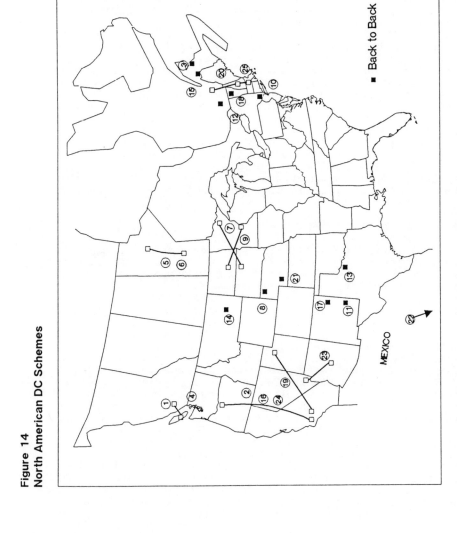

Source: Reeve, "The Location and Characteristics of Recently Completed DC Transmission

Table 19
North American HVDC Schemes in Operation and Active Consideration

	Scheme	Year	Rating (+/- kV)	Rating (MW)	Length of Route (miles)
1	Vancouver Island Pole 1	1968-1969	260	312	43
2	Pacific DC Intertie	1970	400	1440	846
3	Eel River	1972	80	320	Back-to-back
4	Vancouver Island Pole 2	1977-1978	280	370	45.4
2	Pacific DC Intertie	1977	400	1600	846
5	Nelson River 1	1977	450	1620	555
6	Nelson River 2	1977	250	900	575
7	Square Butte	1977	250	500	456
8	David A. Hamil	1977	50	110	Back-to-back
9	CU	1978	400	1000	427
10	Astoria	1981	100-400	100	0.4
11	Eddy County	1984	83	200	Back-to-back
12	Chateauguay	1984	140	500	Back-to-back
13	Oklaunian	1984	83	200	Back-to-back
14	Miles City	1985	83	200	Back-to-back
15	Madawaska	1985	140	350	Back-to-back
16	Pacific Intertie (upgrade)	1985	500	400	846
17	Blackwater	1985	56	200	Back-to-back
18	Highgate	1985	?	150/200	Back-to-back
6	Nelson River 2	1985	500	1800	575
19	Intermountain	1986	500	1600	500
20	Des Cantons-Comerford	1986	450	690	110
21	Walker County	1986	400	520	150
22	Sidney	1988	?	200	Back-to-back
23	Yukatan-Mexico City	?	500	900	?
24	Salt River	1988	314	1600	240
24	Salt River	1988	500	2200	240
25	Pacific Intertie	1988	500	?	846
26	Comerford-Sandy Pond	1990	?	1400	125

Source: Reeve, "The Location and Characteristics of Recently Completed DC Transmission Schemes in North America," 1987.

link has only one conductor, usually at a negative polarity, and the return path is through sea or ground (Gotland DC scheme). The bipolar link has two conductors, one positive and the other negative. Each terminal has two converters of equal voltage in series on the DC sides. The two poles can operate independently if both neutrals are grounded. There is no ground current during normal operation. In the event of a fault on one conductor, the other can operate with return current through the ground (Nelson River DC scheme). The homopolar link has two or more conductors of the same polarity, usually negative, and operates with a ground link (Sardinia DC scheme). Static volt-ampere reactive generators are connected to the AC side of each converter to supply reactive power to the converters. Harmonic filters are connected on the AC side of each converter (Arrillaga 1983, 51). Most of the existing DC schemes comprise two terminals; however, multiterminal HVDC schemes are being planned in the United States and Canada.

SUPERCONDUCTING TRANSMISSION LINES

There has been much interest in the potential of superconducting transmission lines as advances in superconductivity are being reported widely. An increase in critical current density and a technique of making wires from warm superconductors may make economical transfer of large amounts of power over long distances through DC transmission lines that can be placed underground. Underground lines will require less right-of-way and will have less impact on the environment. Most experts believe that the use of warm superconducting transmission lines is at least a decade away. While tremendous progress has been made in raising the temperature of warm superconductors, the theory of superconductivity in these warm superconductors is not very well known at the present time.

A material becomes a superconductor when its electrical resistance drops to zero. Superconductivity was first discovered by a Dutch physicist, Heike Kamerlingh Onnes, in 1911 at 4° K (-452° F). An alloy of niobium and germanium showed superconductivity at 23° K in 1973. In 1986, two IBM scientists, J. George Bednorz and K. Alex Muller, announced a lanthanum copper oxide compound which showed superconductivity at 35° K and won the Nobel Prize for this discovery. Paul Chu and his colleagues at the University of Houston discovered an ytrium-barium-copper oxide ($YB_2Cu_3O_7$) which become a superconductor at 95° K. Many countries have increased their research budget for high-temperature superconducting materials due to the enormous potential economic benefits. In 1988, a team at the National Research Institute for Metals in Tsukuba, Japan, announced a superconducting compound composed of bismuth, calcium, strontium, copper and oxygen which is a superconductor at 90° K. This compound is easier to work with than the Y-B-Cu-O compound.

Low-temperature superconductors require cooling by liquid helium. Warm superconductors can be cooled by liquid nitrogen, which is much less expensive than liquid helium. However, warm superconductors pose tremendous material related problems. These high T_c superconductors are very brittle and have low critical current densities. Computers, electronics, transportation systems, electric motors and generators, transmission lines, and energy storage devices are some of the applications of these superconductors. A low-temperature superconducting magnetic energy storage unit was installed in an electric utility to damp low-frequency oscillations (Boenig and Hauer 1985). Damping of oscillations in power systems allows transfer of more power through transmission lines.

POWER TRADE

An alternative to constructing new generating capacity is to purchase power from other suppliers who may have excess capacity or who are able to produce cheaper electricity. Electric utilities routinely schedule generation and dispatch power so that total operating costs are minimized, given reliability considerations. Such transactions allow utilities to substitute power from less expensive sources of generation, if this power is available, instead of using more expensive power generating facilities. Interest in coal-by-wire arose as eastern utilities sought to substitute less expensive coal-fired produced power for power produced at local oil-fired plants. Transactions between utilities also serve to support utilities when electricity demand is unexpectedly high or when native generation experiences outages.

Bulk power trade has been increasing in the United States. Total generation and sales at major electric utilities more than doubled over the period from 1965 to 1985. Bulk power transactions increased at a much greater rate. Both "interchanges-in" and "interchanges-out" increased more than sixfold over this twenty-year period, as shown in Table 20. There are three major types of bulk power transactions: (1) economy transactions, (2) capacity transactions, and (3) reliability and convenience transactions (U.S. Energy Information Administration 1983). Sales between utilities to reduce costs or to maintain reliability are known as coordination sales. Utilities also make requirement sales to utilities which do not own generating capacity. Economy transactions are those which use lower cost power from another utility instead of some of the purchasing utility's capacity, which is more expensive, although the buyer does not gain additional capacity. In other words, the buyer purchases power which it could have produced. This is normally accomplished through an "economic dispatch" of power in which the dispatch center continuously evaluates available generation that can meet loads, without sacrificing reliability. The incremental cost of generating an additional kilowatt-hour of power from any unit depends upon the nature of the power plant. The dispatch center keeps track of the systemwide incremental cost, known as the "system lambda," to determine the appropriate use of each unit to minimize the total cost (Federal Energy Regulatory Commission 1981, 27). In this way, most large coal, nuclear, and hydroelectric plants, which have lower operating costs, are operated almost continuously as base load units. As demand increases, other "peaking" units which are more expensive to operate are called into service, or power may be brought in from other systems.

Capacity transactions, or firm power transactions, allow a utility to obtain additional generating capacity, for a specified period, from another utility. A utility unable or unwilling to construct additional native capacity can purchase capacity from another utility. Firm power sales are becoming increasingly common. Finally, reliability and convenience transactions are emergency and other transac-

Table 20
Generation, Sales, and Bulk Power Transactions by Major Private Electric Utilities, 1965-1985
(billions of kilowatt-hours)

Year	Generation	Sales to Ultimate Customers	Interchanges In	Out
1965	807.0	730.4	59.6	60.7
1970	1,187.5	1,083.0	161.0	154.6
1975	1,492.6	1,353.0	281.7	254.7
1980	1,784.8	1,620.1	375.4	341.3
1985	1,868.0	1,769.4	412.3	373.2

Note: "Major" private electric utilities are those that have had, in the past three consecutive calendar years, sales or transmission services that exceeded one of the following: 1 million Megawatt-hours of total annual sales, 100 megawatt-hours of annual sales for resale, 500 megawatt-hours of annual gross interchange out, or 500 megawatt-hours of wheeling for others.

Source: U.S. Energy Information Administration, Statistics of Privately Owned Electric Utilities in the United States, Annual.

tions that improve system reliability (U.S. Energy Information Administration 1983).

In response to the Northeast power failure of November 1965, the Federal Power Commission (FPC) investigated the reliability of interconnected systems. In 1967, the FPC recommended that regional joint planning and operation of electric facilities be established (Federal Power Commission 1971, 1-22). Utilities established regional reliability councils and formed the National Electric Reliability Council (NERC), shown in Fig 3. These councils share information to ensure system reliability.

Utilities engage in formal contractual arrangements to control power trade and the price of power. The price of power depends on the nature of the power trade. Emergency transactions to cover, for example, an unplanned outage during a period of heavy demand, is normally exchanged at a rate of 110 percent of the supplier's incremental cost, but with a very high minimum price (Federal Energy Regulatory Commission 1981, 34). Some neighboring utilities have agreements to provide emergency power when needed with the understanding that it will be returned at a later date.

Economy energy transactions are generally covered in bilateral agreements between utilities. These are priced at a mutually agreeable level such as a split-savings formula or one with incremental costs plus a charge factor. The utilities must also monitor the exchanges with respect to existing generating and the system lambda.

A number of power pools, which serve discrete areas, also operate in the United States. Power pools are agreements among utilities to coordinate operations and planning.

There are also "tight" power pools which, in addition to planning resources and operations, actually control operations from a single control center. A tight pool determines the system lambda for all the units in the pool and can dispatch accordingly.

Coal-by-wire trade from the Appalachians to the northeastern United States involves transactions between tight power pools. In the northeastern United States, the New England Power Pool (NEPOOL), the New York Power Pool (NYPP), and the Pennsylvania-New Jersey-Maryland Interconnection (PJM) account for almost every major investor owned utility in the northeastern United States. In the coalfields of western Pennsylvania, West Virginia, Ohio, and Kentucky are two major holding companies which are centrally dispatched and operate as tight power pools. The AEP, headquartered in Columbus, Ohio, has operations in seven states: Indiana, Kentucky, Michigan, Ohio, Tennessee, Virginia, and West Virginia. The Allegheny Power System (APS), headquartered in Greensburg, Pennsylvania, has operations in western Maryland, western Pennsylvania, northern West Virginia, and a small area of Ohio. APS is also centrally dispatched.

In addition to power trade within pools and between adjacent utilities, there has been an increase in power wheeling in the United States. According to a Congressional Research Service study (Kaufman et al. 1987), about 8 percent of electricity generation is transported via wheeling by third parties. Power wheeling is trade between systems that are not interconnected, requiring the use of another system's transmission facilities. For example, AEP is not directly interconnected to Pennsylvania Power and Light, but it could trade power with them by using APS's transmission facilities. Wheeling charges are paid to the utility providing the service; it can be a source of substantial revenue in some cases.

As discussed by Alvin Kaufman et al. (1987), the Federal Energy Regulatory Commission (FERC) has only limited authority to require the wheeling of power. The FERC has not acted to supersede state regulatory authority by granting wheeling and must evaluate wheeling requests on a case-by-case basis. Wheeling is governed by the Federal Power Act, the Atomic Energy Act, and Federal antitrust law, but a request must be made to the FERC for a wheeling order; each request is considered in light of public interest, energy conversation, and reliability. The courts do have the authority to order wheeling because of antitrust considerations, but they defer to the FERC on the conditions of wheeling. The Supreme Court ruled, in the important 1973 <u>Otter Tail Power Company v. Federal Power Commission</u> (precursor of the FERC) decision, that power companies are not immune from antitrust regulation. However, in this and in subsequent court rulings, antitrust considerations alone are not sufficient cause to require wheeling. Mandating wheeling to enhance competition would in essence impose common carrier status on electric utilities. The commission usually approves rates that do not exceed average embedded costs, but it has also been approving market or value based rates for short-term interchanges (Barker 1988).

The issue of whether wheeling should be required for any utility is currently a topic of great debate. Many utilities, and industrial consumers of electricity, favor mandatory wheeling to allow them to "shop around" for the most favorably priced power, bypassing their local utility. Small power producers and some utilities also favor mandatory wheeling to allow them access to lucrative markets. This also impacts decisions regarding Canadian power and coal-by-wire. The ability of Canadian power to be marketed deep within the United States is limited because of the need to obtain wheeling. The clearest example of this was the failure of the Manitoba-to-Nebraska Mandan project which would have connected winter peaking Manitoba and North Dakota to summer peaking South Dakota and Nebraska. Despite demonstrated energy savings to Nebraska, South Dakota blocked the proposed line contending that it would not realize suitable benefits (Kaufman et al. 1987). Coal-by-wire proposals to New England using the existing network would also have the transmission access problem and may also be faced with multiple wheeling charges as the power passes through several utility jurisdictions, making the power less competitive.

In March 1988, the FERC proposed changes relating to the Public Utilities Regulatory Policy Act (PURPA) dealing with competitive bidding, independent power producers, and the determination of avoided cost (Romo 1988). These changes will move the industry to a more competitive environment and will relax regulations for independent power producers (IPPs). However, the transmission access problem will still be outstanding. As argued by the proponents of mandatory wheeling, competitive bidding can not be truly competitive unless nonadjacent generators are allowed to compete in the market. In order to accomplish this, it is necessary to have greater access to transmission facilities.

Because electricity cannot be directed in the interconnected AC system, there arises "looping flow" or "inadvertent flow" whereby power passes through systems not involved in the transaction or benefitting from wheeling charges. In the early 1980s, when oil prices were extremely high, this feature of AC transmission was widely discussed because this looping flow was seen to be interfering with the west-to-east trade in economy energy. Fig 15 illustrates the situation. In this case, there is a contract between Ontario Hydro and the NYPP for the sale of 1,000-megawatts of power to New York. According to a study conducted by the utilities in the region, this 1,000-megawatts would actually flow through many paths according to the electrical characteristics of the network, and not just directly between Ontario Hydro and the NYPP. (ECAR/MAAC Coordinating Group 1985, 6). About half of the power would flow directly to New York from Ontario; the other 500-megawatts would circulate through Michigan and cross Pennsylvania, predominantly through the APS. Partly as a result of this, the Mid-Atlantic Area Reliability Council (MAAC) area's transmission system was about 97

Figure 15
Interconnected System Response for Ontario Hydro to NYPP 1000 Megawatt Schedule (response in megawatts)

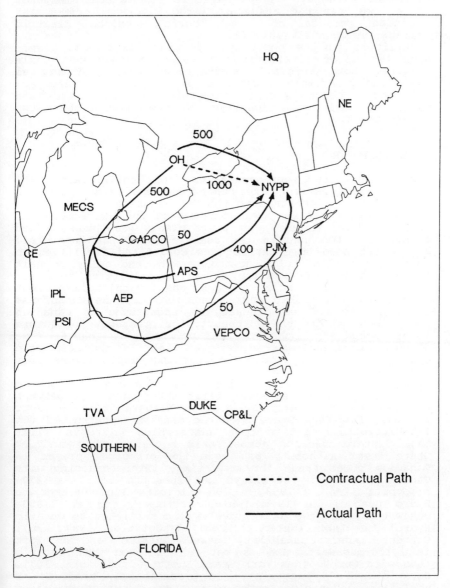

Source: ECAR/MAAC Coordinating Group, <u>ECAR/MAAC Interregional Power Transfer Analysis</u>, 1985.

percent utilized through 1983 and 1984 (ECAR/MAAC Coordinating Group 1985, 4), limiting trade from the West.

The problem in this region is partly a result of using the interconnected system for a purpose for which it was not constructed. Interconnections between systems grew mainly to provide reliable service to a utility's own customers. The system was built to accommodate emergency trade, not for the trade of large blocks of power for extended periods of time (North American Electric Reliability Council 1984, 4).

Utilities in the Northeast, seeking lower cost power, have two major alternatives: Canadian power and coal-by-wire. Both sources of power are produced by economical base load generators. Canadian power includes hydroelectric power, which is very inexpensive, nuclear power, and even some coal-fired capacity. We now discuss these two alternatives.

CANADIAN POWER

Coal industry representatives and politicians from coal-producing states are concerned about the growth in Canadian power imports into the United States. Concern is voiced that Canadian power may reduce the utilization in existing coal-fired generating plants in head-to-head competition (whether in the coalfields or load centers), reduce the need to construct new coal-fired facilities to meet future electricity demand growth, and delay or halt the conversion of existing oil-fired facilities to coal. In all cases, coal sales (present and future) will be lost to the growing market share controlled by the Canadians. In this section, we discuss the background on Canadian power imports into the Northeast, the prospects for future imports, and its impact on coal-by-wire. In Chapter 7 we discuss Canadian power imports in light of the acid rain issue, which has some far-reaching implications for coal-by-wire.

There has been electricity trade between the United States and Canada since the first interconnection between the two countries was completed at Niagara Falls in 1901. In 1942, New York State Electric and Gas constructed two 115-kilovolts interties with Canada which marked the first major interconnection across this border. The New York State Power Authority has been the principal player in Canadian power trade for some time, interconnecting with Canada with two 230-kilovolts lines in 1955, a 230-kilovolts line in 1958, a 765-kilovolts line (linked to Hydro Quebec) in 1978 (which is now rated at 2,350-megawatts), and two 345-kilovolts lines (linked to Ontario Hydro) in 1981 (National Coal Association 1987, 26). Current transfer capability between Ontario and New York is 2,150-megawatts and is better connected with Ontario than with Quebec (New York State Energy Office 1985, 107). Ontario Hydro's electricity is predominately produced at coal-fired and nuclear power plants; the power generated by Hydro Quebec is hydroelectric. Although New York does have transmission interconnections with Canada, New York State's need is to displace the power generated in the New York

City area's oil-fired plants. The New York Power Authority has had problems constructing the 345-kilovolts transmission line from upstate New York to the lower Hudson River valley. The completion of this project will allow an additional 2,500-megawatts of power to move across New York State (Hiney 1986, 124-25).

Although there has been a long history of power trade between New England and Canada, the major interconnections were not completed until recently. In 1969, a 345-kilovolts line was completed interconnecting New Brunswick to NEPOOL, and in 1986 a 450-kilovolts (DC) line interconnecting Hydro Quebec to New Hampshire was completed. This line, rated at 690-megawatts is to be upgraded to 2,000-megawatts by the early 1990s (National Coal Association 1987, 26). In addition to this line, New England is connected to New Brunswick by a 345-kilovolts, a 120-kilovolts and five 69-kilovolts lines and by two 120-kilovolts lines from Vermont to Quebec. Fig 16 shows the location of the major interconnections to Canada from the Northeastern United States.

Only recently have there been substantial imports of Canadian power into the United States. In the early 1970s, electricity trade between the two countries was well balanced. As shown in Table 21, 9,214 million kilowatt-hours of net Canadian power were brought into the United States in 1976; by 1985, net Canadian power imports exceeded 40,000 million kilowatt-hours. This transaction accounts for about 2 percent of United States electricity sales and about one-tenth of the total electricity generation in Canada (U.S. Department of Energy 1987, 2). For New York and New England, however, Canadian power is making a significant contribution to the total electricity supplies.

It is not the absolute level of Canadian power imports that is so troublesome to coal interests, but the fact that most of the imports are to regions where coal has large potential markets. About 60 percent of the total U.S. imports of Canadian power are purchased by New York and New England. In 1985, 23,962 million kilowatt-hours went to New England, New York, and New Jersey. In New York State, for instance, in 1985 up to 17 percent of the state's electricity requirements were being provided by Canada, compared to only 2 percent in 1977 (New York State Energy Office 1985, 104). New England imported 6,675 million kilowatt-hours of electricity in 1985, almost all from New Brunswick. These imports originated primarily from the Lepreau 1 nuclear plant which was completed in 1983.

The rationale for increasing Canadian electricity imports into the Northeast is quite compelling. Canada has vast hydroelectric potential, the cost of generating the power is low, and the systems match well in terms of load characteristics. Canadian utilities are winter peaking whereas the U.S. utilities are summer peaking, thus electricity trade across the international boundary reduces the total amount of capacity required. The case is vividly portrayed by the summer and winter reserve margins of the Northeast Power Coordinating Council (NPCC), which comprises twenty-two systems in the northeastern United States and eastern Canada. In the summer, the reserve

Figure 16
Major U.S.-Canadian Transmission Line Interconnections, 1985
Northeast United States, Ontario, Quebec, and New Brunswick:
(69 kilovolts and larger)

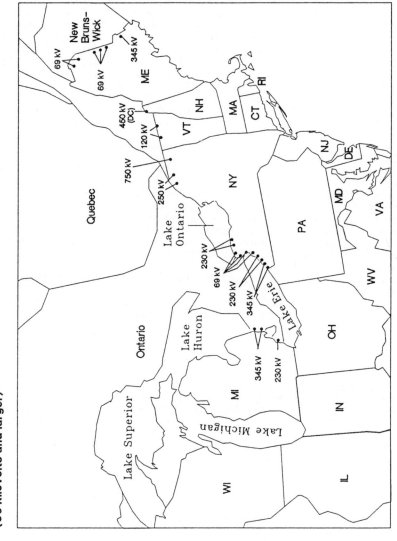

Source: U.S. Energy Information Administration, U.S.-International Electricity Trade, 1986.

Table 21
Net U.S. Imports Of Canadian Power (million kilowatt-hours)

Year	Net U.S. Imports
1976*	9,214
1981**	33,899
1985**	40,843

Source: * U.S. Energy Information Administration, U.S. Canadian Electricity Trade, 1982.

** U.S. Energy Information Administration, U.S. International Energy Trade, 1986.

margin for 1985 in the United States portion was 28.3 percent compared to 79.5 percent in the Canadian portion. In winter 1985, the U.S. reserve margin was 34.2 percent compared to 20.2 percent for Canada (National Coal Association 1987, 26).

Canada boasts great potential for further hydroelectric development. Table 22 summarizes the peak capacity of existing, planned, and potential hydroelectric sites across the Canadian provinces. This table does not include the capacity of nuclear or coal-fired facilities. Currently, about 55,000-megawatts of peak capacity are available at existing hydroelectric stations. Another 9,371-megawatts are now under development, with a potential of another 63,000-megawatts at undeveloped sites. As shown in the table, Quebec, which has the greatest hydroelectric resource base, could bring forth up to another 34,000-megawatts of capacity. Not only is the province well-endowed, but the premier of Quebec, Robert Bourassa, has outlined a strategy in his book Power from the North to satisfy Quebec's electricity needs and to bring forth an additional 12,000-megawatts for export to the United States by the year 2000.

The cost differential for power produced in Canada compared to that produced in the northeast United States is also large enough to encourage the development of further electricity trade. The U.S. Department of Energy concluded that over 90 percent of Quebec's undeveloped hydroelectric resource would be less expensive than coal-fired electricity even if it was financed with a U.S. private capital structure under the U.S. tax system (U.S. Department of Energy 1987, 11). The La Grande project in Quebec actually cost less than originally anticipated. The initial estimate for the project was a cost of $17.2 billion (1985 dollars) and a production of 8,300-megawatts; the final project cost was $16.5 billion with a capacity of 10,282-megawatts (U.S. Department of Energy 1987, A-9).

There are two problems with Canadian power. One is the concern over reliability and the other is the effect power purchases on Canada will have on the nation's balance of payments. The reliability question has two components.

Table 22
Existing, Planned, and Undeveloped Hydroelectric Sites in Canada (peak megawatts)

Canadian Province(s)	Existing at End of 1984	Near-Term Planned Additions	Undeveloped Sites in Study	Total in Table
Newfoundland & Labrador	6,213	120	2,300	8,633
Maritime	1,269	160	0	1,429
Quebec	24,877	2,880	34,729	62,486
Ontario	7,130	1,067	0	8,197
Manitoba	3,641	2,920	3,096	9,651
Saskatchewan	575	0	0	575
Alberta	734	0	1,600	2,334
B. Columbia	10,379	2,224	21,750	34,353
Yukon & NW Territories	130	0	0	130
Total	54,948	9,371	63,475	127,794

Sources: Data on existing sites from 1984 year-end census by Statistics Canada; data on planned and undeveloped sites per correspondence with Hydro Quebec, Manitoba Hydro, and British Columbia Hydro from U.S. Department of Energy, Northern Lights, 1987.

The first is the ability of Canada to maintain a reliable supply of power without interruption. On Monday, April 18, 1988, a failure in a substation in northern Quebec, caused by a snowstorm covering transformers, condensers, and switches in ice, cut off electricity to about 6 million people in Quebec and parts of New England (Perley 1988). The province experienced a similar blackout in 1982 when a transformer exploded at a substation. Officials in the United States are concerned about the ability of Canada to provide reliable power that will not jeopardize their own system's reliability. William Sheperdson of the New England Power Pool said that Americans cannot tolerate Hydro Quebec's frequent outages (Toronto Star 1988, A-17).

The second reliability concern is that Canada may not sell power to the United States if it needs the power to meet its own electricity requirements. This fear has been ameliorated somewhat by the U.S.-Canada Trade Agreement. According to the agreement, if an energy supply problem occurs, available energy will be shared proportionately between domestic and export markets and there will be no quotas. The U.S.Canada trade agreement will be discussed in more detail in Chapter 7.

COAL-BY-WIRE

The construction of coal-fired power plants close to the coal mines became increasingly desirable throughout the 1960s and 1970s. The Virginia Electric Power Company (VEPCO) and General Public Utilities were leaders in this concept. General Public Utilities, which served customers in Pennsylvania and New Jersey, constructed the Keystone and Conemaugh plants in western Pennsylvania in the late 1960s and early 1970s, which generated power for customers in the East. VEPCO constructed the Mt. Storm plant in a rural area in West Virginia in the mid-1960s and had the power transmitted to customers out of state.

Over the 1963-70 period, the electric utility industry completed construction of over 18,000-megawatts of mine-mouth coal-fired power plants, from units 500-megawatts or greater. Over half of this capacity (9,567-megawatts) was sited in western Pennsylvania, Ohio, or West Virginia. From 1971 to 1975, another 18,000-megawatts of mine-mouth power was completed from units at least 500-megawatts in size, and almost 60 percent of this capacity (10,755-megawatts) was sited in western Pennsylvania, Ohio, or West Virginia (Federal Power Commission 1971, 4-13). Some of these plants, such as Keystone, Conemaugh, and Mount Storm, were designed to serve out of state customers, but most were built to serve local demand. It was projected in 1970, for instance, that the peak electricity loads in the East Central Area Reliability Council (ECAR) region by 1980 would be 81,900-megawatts (Federal Power Commission 1971, 3-14); however, by 1983, the load in the region totaled only 63,800 megawatts (North American Electric Reliability Council 1984, 16).

The growing price disparity between oil and coal after 1970 has encouraged the displacement of oil by electricity generated in the coalfields. Fig 17 shows the trend in the coal-oil price ratio over the period from 1969 to 1985. A ratio of 1 means that the delivered price of coal equals the delivered price of oil. In 1969, the coal-oil price ratio was about 0.80 nationally, meaning that coal was slightly less expensive to use than oil. In New England, the coal-oil price ratio exceeded 1.25 so that oil enjoyed an advantage over coal delivered to the region. The coal-oil price ratio declined steadily, then dropped rapidly in response to the oil price rises in 1973-74. The ratio fell below 0.5 nationally, providing the economic advantage to coal. The rising price of oil encouraged power trade from the mine-mouth plants of the ECAR region. In the early 1980s, electricity from the ECAR region transmitted to the MAAC region was displacing approximately 40 million barrels of oil per year (ECAR/MAAC Coordinating Group 1985, 2). Electricity trade from ECAR to MAAC averaged about 2,300-megawatts per day. The coal-fired power plants were not built to engage in power trade, nor was the electricity transmission network designed for this level of economy energy transfers.

In the early 1980s, the system was transferring about as much power as the network could accommodate. At some times, up to 3900-megawatts of economic transfer was occurring per day. New England has also expressed an interest in considering purchasing economy power, but the transmission constraints make it too difficult to serve the region with power from the coalfields. Improvements are being made to the system which will allow more power to be transferred from the West to the East. Phase shifter controls, for instance, are being added to reduce the inadvertent flow of power, and the completion of the

Figure 17
Coal–Oil Price Ratio, 1969–1985

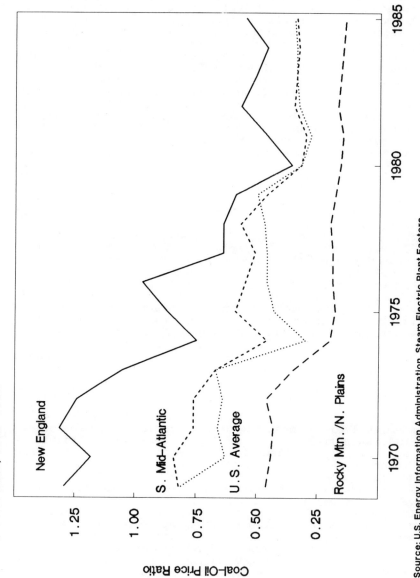

Source: U.S. Energy Information Administration, Steam Electric Plant Factors, 1969–1985.

transmission loop around Washington, D.C., should facilitate greater transfer of power.

It is probably necessary to build new high-voltage transmission lines in order to expand coal-by-wire trade particularly for firm power contracts. New "conceptual plans" involve the construction of new transmission facilities from ECAR to MAAC and even to NEPOOL. These plans include a 765-kilovolts line from a terminal in the northern West Virginia panhandle to Three Mile Island,-a series-compensated 500-kilovolts line connecting the same points that would be compatible with existing 500-kilovolts voltage networks, and 500-kilovolts high-voltage DC lines which would require converter/inverter facilities at each terminal (ECAR/MAAC Coordinating Group 1985, 1). These lines, which would cost from $500,000,000 to $1,000,000,000 to build, would be able to improve transfer capability by from 1,500 to 2,500-megawatts. As of this date, no plans have been announced to construct new transmission lines.

CONCLUSION

The nation's electricity transmission system is highly complex and involves close coordination among the nation's electric utilities. The system was designed first to serve local areas, but it evolved into a regional system as neighboring utilities coordinated their operations for reliability and economy purposes. However, the system as it evolved, was not designed to facilitate large-scale interregional bulk power transfers on a continuous basis. The system has been handling large-scale economy energy transfers in the East, and there is an interest in adding new facilities to increase the power trade options.

It is generally recognized that there is a need to address the electricity transmission question in the United States. The National Governors' Association transmission task force is a good vehicle for this. The broad participation of the nation's governors in this study indicates the concern with the transmission problem. In general, the governors have found that opportunities for power trade, and substantial savings to customers in electricity-exporting and importing regions, could be achieved by expanding the transmission network.

We will return to this discussion, but first we wish to investigate whether electricity exporting regions indeed benefit by the sale of electricity. We consider both the construction of new facilities and the utilization of existing capacity on a regional economy. We will demonstrate these impacts by investigating effects to the state of West Virginia.

BIBLIOGRAPHY

Arrillaga, J. 1983. High Voltage Direct Current Transmission. Peter Peregrinus, Ltd.

Barker, James V. 1988. "A Workable Test of a Workably Competitive Bulk Power Market." Public Utilities Fortnightly 121, 8 (April): 13-77.

Boenig, H. J., and J. F. Hauer. 1985. "Commissioning Tests of the Bonneville Power Administration 30 MJ Superconducting Magnetic Energy Storage Unit." IEEE Transactions PAS-104 (February).

Dunlop, R. D., R. Gutman, and P. P. Marchenko. 1979. "Analytical Development of Loadability Characteristics for EHV and UHV Transmission Lines." IEEE Transaction on Power Apparatus and Systems 98(March/April): 606-17.

ECAR/MAAC Coordinating Group. 1985. ECAR/MAAC Interregional Power Transfer Analysis. Report presented to the U.S. Department of Energy, June 1985.

Federal Energy Regulatory Commission. 1981. Power Pooling in the United States. Washington, D.C.: Federal Energy Regulatory Commission.

Federal Power Commission. 1971. The 1970 National Power Survey Part 1. Washington D.C.: U.S. Government Printing Office.

Hiney, Robert A. 1986. "Importing Canadian Electricity." In The Future of Electrical Energy, edited by S. Saltzman and R. E. Schuler, 123-25. New York: Praeger.

Hingorani, Narain G. 1986. "Emerging Technical Solutions to Electric Power Transmissions" Paper presented at Northeast-Midwest Institute Forum, Itasca, Ill., May 1986.

Kaufman, Alvin, Carl Behrens, Donald Dulchinos, Larry B. Parker, and Robert D. Poling. February 12, 1987. Wheeling in the Electric Utility Industry Washington, D.C.: Congressional Research Service.

Kimbark, Edward W. 1971. Direct Current Transmission, New York: John Wiley & Sons.

National Coal Association. July 1987. The Effects of Imports of Canadian Electricity on the U.S. Coal Industry, 1980-85. Washington, D.C.: National Coal Association, Policy and Analysis Department.

New York State Energy Office. 1985. Meeting the Challenge: An Analysis of Electricity Supply Options for New York State. Albany: New York State Energy Office.

North American Electric Reliability Council. 1984-1985. Electric Power Supply and Demand. Princeton, N.J.: North American Electric Reliability Council, 167.

---. 1984. Impediments to Transfers. Princeton, N.J.: North American Electric Reliability Council.

Perley, Warren. April 19, 1988. "Two Massive Power Failures Blackout Quebec, Montreal." United Press International.

Reeve, John. 1984. "The Location and Characteristics of Recently Completed DC Transmission Schemes in North America." Paper presented at the International Conference on DC Power Transmission, 43-47.

Romo, Cheryl. 1988. "Observers Predict Electric Proposals Will 'Fundamentally Change' Industry." Public Utilities Fortnightly 121, 8 (April 14,): 40-44.

Toronto Star. April 21, 1988. "U.S. Official Criticizes Hydro-Quebec Service," A-17.

U.S. Department of Energy. December 1987. Northern Lights: The Economic and Practical Potential of Imported

Power from Canada. Washington, D.C.: Office of Policy, Planning and Analysis.

U.S. Energy Information Administration. 1986. *U.S. International Electrical Trade*. Washington, D.C.: U.S. Department of Energy/U.S. Energy Information Administration.

---. 1983. *Interutility Bulk Power Transactions*. Washington, D.C.: U.S. Department of Energy/U.S. Energy Information Administration.

---. 1982. *U.S. Canadian Electricity Trade*. Washington, D.C.: U.S. Department of Energy/U.S. Energy Information Administration.

6
Economic Impacts of the Electricity Industry in West Virginia

Frank J. Calzonetti, Gregory G. Sayre, John Merrifield, and Tom S. Witt

The economic impact of the electric power industry in any coal-producing state is substantial, and West Virginia's experience is no exception. Attempts to quantify the impact may fail to identify all aspects of the impact. Although this chapter develops several measures of the economic impact on the West Virginia economy, it is highly possible that certain economic impacts will be left unmeasured. Recognizing the limitations of economic impact analysis at the state level, this study utilizes a conservative methodology which, if anything, will tend to underestimate the economic impact of electric utilities on the state economy.

Attention first is focused on the historical linkage of coal with the electric utility industry. Patterns of coal production and consumption in West Virginia are traced and compared with similar national patterns. Also examined are similar trends in coal-mining employment, wages, and personal income during the same time period.

The actual consumption of coal by electric utilities and the resulting electricity production in West Virginia are examined. Following this is an examination of the electricity exports from West Virginia and the determination of their economic impact on the state economy and local communities. The chapter also evaluates the potential of abundant and inexpensive electricity to attract industry to the state.

STATEWIDE ECONOMIC IMPACT OF POWER PLANTS

The principal analytical tools utilized to measure the economic impact are the various multipliers derived from the most recent version of the West Virginia Input-Output Model (Greenstreet 1987). The various multipliers are defined and are used to assess the total economic impact of the electric utility industry in West Virginia. In addition, the methodology allows for an assessment of the impact of a single plant and varying levels of operation of a plant on the state economy.

The conservative nature of the economic impact measures is partially attributable to the conservative multiplier estimates. In addition, the study does not estimate several quantitative and qualitative impacts of the electric utility industry on the state economy. First, unlike many businesses and households, electricity generation and distribution systems require minimal amounts of governmental services. Second, electric utilities account for a minimal amount of the tax collection costs incurred by governmental units. Third, employment patterns within electric utilities are reasonably stable over time, particularly when compared with other sectors. Finally, electric utility employees pay a variety of state and local taxes, a portion of which are mentioned in this chapter.

West Virginia was the third largest producer of coal in the United States in 1985 with an annual production that exceeded 127 million tons. The value of this coal was approximately $4.2 billion (West Virginia Coal Association 1986). The industry employed about 32,000 coal miners for the year at an average weekly wage of $628.00 (West Virginia Coal Association 1986); however, the number employed has declined since 1985 in part due to further introductions of long wall mining technology and declining real energy prices. Over three-fourths of the West Virginia coal shipped domestically was purchased by electric utilities. About 22 percent of the coal shipped to electric utilities was sent to West Virginia power plants. Thus, West Virginia power plants represent about 18 percent of the total market for the West Virginia coal distributed in the United States, and about 13.2 percent of the total coal shipped including exports (see Table 23). The electric utility industry is very important for the West Virginia coal industry.

These figures, however, underestimate the contribution of West Virginia coal to West Virginia power plants. Table 24 shows that the state's thirteen major power plants consumed 27.8 million tons of coal in 1985 of which 26.011 million tons was identified as originating in West Virginia. Fig 18 shows the location of these plants in West Virginia. The apparent divergence in the amount of coal produced and delivered to West Virginia power plants (26.011 million tons versus 17.093 million tons in Table 23) is attributable to the multiple ownership of West Virginia power plants. The portion of the coal delivered to a multiple ownership power plant attributable to a West Virginia utility is included in both sets of statistics. The portion of the coal delivered to the same plant attributable to a non-West Virginia utility is excluded from the 17.093 million ton figure and is included in the 26.011 million ton figure. As a result, the more accurate measure of the amount of West Virginia coal delivered to power plants located in West Virginia is the larger amount, 26.011 million tons which represents 20.1 percent of all coal production in West Virginia.

Assuming that 20.1 percent of all coal-mining employment was directly attributable to shipments to West Virginia electric power plants, it is estimated that 6,490 employees

Table 23
Distribution of West Virginia Coal, 1985

	Thousands of Tons	Percent
West Virginia Electric Utilities	17,093	13.2
Electric Utilities in Other States	55,416	42.8
Electric Utilities in Canada	4,244	3.3
Coke Plants	11,741	9.1
Other Industrial	9,918	7.7
Residential/Commercial	782	0.6
Exports (not included above)	30,089	23.2
Unknown	109	0.0
Total	129,392	100.0

Note: Totals may not correspond due to rounding. In addition, coal distributed includes some mined in late 1984 and omits coal mined in late 1985.

Source: West Virginia Coal Association, *Coal Facts 1985, 1986*.

were directly associated with the production of coal for shipment to these plants. Additional employment was also associated with the transportation and allied areas, making the figure in excess of 6,500 employees directly associated with the production and distribution of West Virginia coal to West Virginia power plants.

In 1985, the total installed generating capacity of these thirteen coal-fired power plants was 14,957 megawatts, net generation was 79,168,000 megawatt-hours (National Coal Association 1986). The overall capacity factor, as shown in Table 25, was 60.3 in 1985. In general, the electricity sector in West Virginia was operating somewhat below the 80-percent rate, which is considered to reflect the full utilization of a power plant, taking into account scheduled maintenance, reserve margin requirements, and unplanned outages.

The thirteen power plants in 1985 directly employed 3,095 individuals and consumed $1.1 billion worth of coal. Over 90 percent of the coal consumed in these plants originated at West Virginia coal mines. Even though these power plants employ relatively few workers, the contribution of these plants to the West Virginia economy is enormous. In addition to the quantity of coal purchased and the workers employed, given that about 70 percent of the electricity from these plants is sold outside West Virginia, electricity exports contribute over $2 billion annually to the West Virginia economy. The industry also

Table 24
Coal Consumed at West Virginia Power Plants, 1985

Power Plant	Owner	Total Coal Consumed (million tons)	Percent of Coal Produced in West Virginia
John Amos	Appalachian Power Ohio Power	5.432	100.0
Kanawha	Appalachian Power	0.540	100.0
Mountaineer	Appalachian Power	2.122	100.0
Philip Sporn	Appalachian Power	1.290	73.0
Albright	Monongahela Power Potomac Edison	0.633	83.0
Fort Martin	Monongahela Power Potomac Edison West Penn Duquesne Light	2.324	64.5
Harrison	Monongahela Power Potomac Edison West Penn	3.863	100.0
Pleasants	Monongahela Power Potomac Edison West Penn	2.818	96.5
Rivesville	Monongahela Power	0.187	97.7
Willow Island	Monongahela Power	0.338	17.7
Kammer	Ohio Power	1.881	100.0
Mitchell	Ohio Power	3.011	100.0
Mount Storm	Virginia Electric and Power	3.382	96.9

Source: West Virginia Coal Association, Coal Facts 1985, 1986.

paid considerable taxes to state and local economies, as shown in Table 26.

Recent work has resulted in the first major revision of the West Virginia Input-Output Model since the 1975 version (Greenstreet 1987). This 1982 model is based on the 1977 Bureau of Economic Analysis (BEA) National Input-Output Model and Tables, the 1977 Commodity Transportation Survey, the 1982 Economic Census, and other 1982 data from various governmental sources. Within this model, a variety of economic multipliers are available to assess the economic impact on the West Virginia economy of changes in the electric power industry (sector 399).

Figure 18
Electric Generation Plants in West Virginia

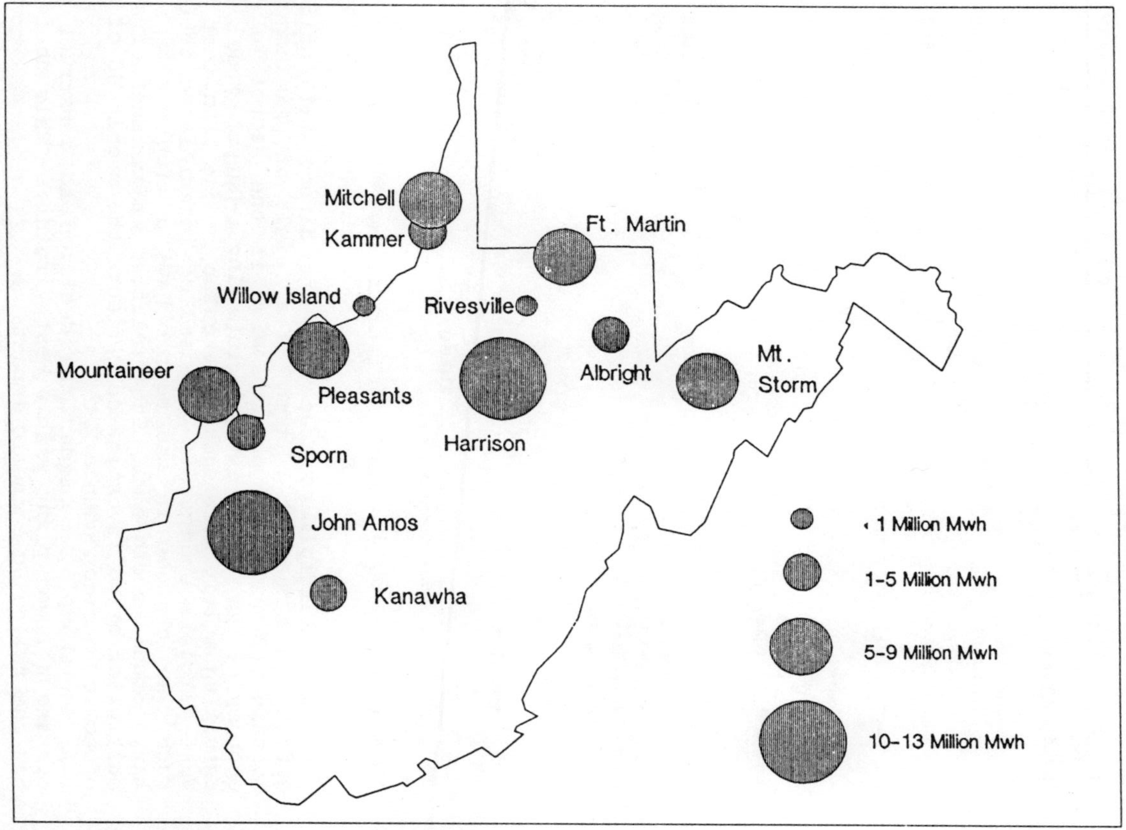

Source: West Virginia Public Service Commission.

Table 25
Installed Capacity and Capacity Factors at West Virginia Power Plants

Owner and Plant	Capacity MW	1985 Capacity Factor
Appalachian Power Co. (AEP)		
John E. Amos	2,932.6	67.4
Kanawha River	439.4	35.8
Mountaineer	1,300.0	62.1
Philip Sporn*	1,105.0	43.2
Ohio Power Co. (AEP)		
Kammer	712.5	74.3
Mitchell	1,632.0	59.1
Monongahela Power Co.		
Albright	278.0	52.9
Ft. Martin	1,152.0	64.9
Harrison	2,052.0	60.8
Pleasants	1,368.0	59.7
Rivesville	109.0	41.1
Willow Island	215.0	47.8
Virginia Power Co.		
Mount Storm	1,662.4	60.9
TOTAL	14,957.9	60.3

Note: *Also owned by Central Operating Company

Source: National Coal Association, Steam Electric Plant Factors 1986, 1986.

Economic multipliers trace out the direct and indirect effects on a subject economy as a result of permanent or sustained contraction or expansion in one sector of the economy. The 1982 West Virginia Input-Output Model and resulting multipliers incorporate the recognition that the West Virginia economy is composed of interrelated sectors such that changes in one sector have a ripple effect on many other sectors. The relative magnitude of the multiplier being investigated reflects the magnitude of the economic interdependence.

As an example, consider the decision to construct and operate a power plant within West Virginia. This decision has the following economic impacts:

a. During the construction phase of the power plant, labor, materials, and equipment are purchased by the contractors. Suppliers of these resources find employment and income opportunities created by the

Table 26
Electric Utility State and Local Taxes in West Virginia, 1986

Electric Utility	Real and Personal Property	State B & O Taxes	Municipal B & O Taxes	PSC Fees	Income
Appalachian Power Co.	$8,974,292	$20,230,398	$4,430,051	$1,182,713	$3,791,796
Monongahela Power Co.	$3,506,936	$11,397,988	$3,027,741	na	$1,661,221
Potomac Edison Co.	$4,248,964	$ 7,093,439	$ 327,871	na	$ 448,344
Virginia Electric Power Co.	$2,836,691	$10,058,959	$ 67,242	$ 198,932	$ 782,000

Notes: na -- not applicable.

Table does not include minor tax categories or taxes paid in West Virginia by Ohio Power Co., Duquesne Light Co., and West Penn Power Co.

Source: Data collected by the authors from the West Virginia Public Service Commission.

construction project. Indirect economic effects occur when the direct beneficiaries of the construction project purchase goods and services in the local economy. The extent to which leakages occur outside the local economy will affect the magnitude of the indirect multiplier. Upon completion of the construction project, the income and employment impacts diminish to zero; consequently, the economic multiplier effects associated with the construction phase are temporary in nature. Although temporary, these effects can be sizeable and can result in a substantial economic impact on the local economy.

b. When the completed electric power plant enters service, the operation of the plant results in sizeable purchases of coal, labor, and other resources. Suppliers of these resources have employment and income opportunities created by the generation of electric power from this plant. Indirect economic impacts occur when the direct beneficiaries purchase goods and services in the local economy. These multiplier effects are relatively permanent in nature and have maximum impact on the local economy only if the generation facility is operated at its maximum generation capacity. If the generation facility reduces its scale of operation or is taken out of service, then the multiplier effects operate in the opposite direction leading to indirect impacts such as layoffs and coal mine shutdowns.

In this chapter, use is made of both payroll (similar to income) and employment multipliers. Within each multiplier category, one can distinguish between Type I and Type II multipliers (Miernyk 1965; Greenstreet 1987). Type I multipliers measure the direct and indirect changes in either payroll or employment due to an increase of one dollar's worth of sales by the electric utility industry. Type II multipliers estimate the direct and indirect, as well as the induced changes, in either payroll or employment due to an increase of one dollar's worth of sales to final consumers.

For example, if the Type I payroll multiplier for the electricity sector is 1.464, a $1.0 million increase in the demand for electricity produced in West Virginia will result in a $1.464 million increase in the state payroll. The multiplier effects on the state economy occur irrespective of whether the increase in the sales of electricity is within or outside of the state.

For this study, economic multipliers from the 1982 West Virginia Input-Output Model were utilized. The estimated payroll multipliers were 1.464 for Type I and 1.579 for Type II; the employment multipliers were 2.573 for Type I and 3.503 for Type II.

It should be noted that input-output models have some limitations. Underlying the input-output model is a fixed coefficient production function which implies that past relationships between inputs and outputs from the production process will continue in the future. To the extent that new technology is incorporated in new power plants, which results in a substantially different technical relationship between inputs and outputs, the resulting payroll and employment multipliers may be misleading in measuring the economic impacts of the plant on the local economy. As a result, the multiplier estimates should be viewed only as indicative of the effects on the local economy of the electric utility sector operation. In this context, the estimate of the payroll effect has less uncertainty attached to it than the estimate of the employment effect. The multiplier effect on employment takes much longer to work out in the economic system, and it relies on the critical assumption of no underutilization of labor in the sectors of the West Virginia economy which interact with the electric power sector. This assumption, however, may be more unrealistic today given the predominant technology employed in mining coal underground, long wall mining, which can generate large increases in coal production with minimal changes in employment.

The economic multipliers can be utilized to examine the economic effects in West Virginia of different utilization rates of a coal-fired generation plant similar to those currently being considered for siting in the state. The economic impact of the construction and operation of this plant could be considered along with the economic impacts of operating the plant at various levels of capacity. Although such a plant is hypothetical, the results could indicate the potential economic impacts of varying utilization rates. Normal utilization rates for power plants peak around 80 percent, which is considered to be full utilization due to reserve margin requirements, scheduled maintenance, and planned outage. There are instances, however, when power plants operate continuously for extended periods of time but still may fall below full utilization. Recently, the Appalachian Power Company's

Mountaineer Plant operated continuously for over 400 days, which set a new world record for continuous operation of a power plant.

For the purposes of this study, however, the economic impacts of an individual plant were not considered; instead, the impacts of the entire electric power generation sector on the West Virginia economy under different rates of capacity utilization were examined using the above multipliers. The increase in the utilization rate would most likely occur as a result of an increased demand for electricity exports. The base line data for the electricity sector in 1985 is employed to generate the potential economic impacts from increasing the capacity utilization rate for the electricity sector. In generating these effects, use is made of the Type I payroll and employment multipliers cited above.

Assume the utilization rate of the entire sector is increased to 80 percent from the actual 1985 level of 60.3 percent (Table 25). The 105,032,000 megawatt-hours generated at this higher utilization rate would result in annual sales of $3833.7 million (utilizing 3.65 cents/kilowatt-hour), an increase of $944 million from the actual level. The coal required by the plants is now estimated at 34.509 million tons, an increase of 8.498 million tons. The estimated employment in coal mining associated with this higher utilization rate is 8,626, an increase of 2,126 employees.

The above effects represent only the initial effect on the electric utilities and the direct effect on coal mining from increasing the utilization rate to 80 percent. Use of payroll and employment multipliers permit an estimate of the indirect and induced effects on the state economy from the higher production level. The conservative payroll effect is an increase in annual state payroll of $1490.6 million (Type II payroll multiplier of 1.579 applied to the increased sales of $944.0 million). This payroll increase will result in increased tax revenues (not calculated). The conservative employment effect, an increase of 2,429 permanent jobs within the state, essentially involves the cumulative total of new jobs after the multiplier effects of the sales increase have been realized.

It should be noted that electricity consumers in West Virginia can also benefit from the operation of the power plant system at a higher utilization rate. If the off-system or bulk power sales of electricity generated in West Virginia are at prices exceeding the incremental cost of generation, then the regulatory process (the West Virginia Public Service Commission) can provide that the revenue requirement of the electric utilities regarding in-state retail customers are adjusted downward accordingly. Thus, the electricity consumer in West Virginia can generally benefit from bulk power or off-system sales to consumers in other states, which will provide an additional stimulus to economic development.

This analysis clearly indicates that the impacts of the entire electricity sector in West Virginia operating at different utilization rates are quite substantial and can represent a significant economic impact. The quantified impacts are viewed as conservative estimates since they utilize the most conservative multiplier effects associated with the West Virginia Input-Output Model. In addition, since they measure only the economic impacts that are quantifiable, such estimates ignore many qualitative dimensions associated with economic development.

IMPACT OF GENERATING FACILITIES ON LOCAL AREAS

For the areas in which the construction of large-scale electric generating facilities will occur, the socioeconomic impact of such facilities is often a major community concern. As opposition to many energy developments increased in the 1970s and 1980s and as environmental regulations limited the allowable air-pollution emissions in urban areas, power plants began to be sited increasingly in rural counties.

Rural areas afford the energy developer a number of siting advantages. As mentioned earlier, air quality standards could be more readily met in such areas than in more populated, industrialized centers. In addition, community attitudes toward energy developments are more positive in rural areas. Rural residents tend to favor all types of development more than urban dwellers (Murdock and Leistritz 1979; Bachtel and Molnar 1981; Stout-Wiegand and Trent 1983). Environmental and community groups which often oppose such developments tend to be more organized and better equipped to block such developments in urban areas or more industrialized states. Land costs associated with site acquisition are generally less in rural areas. Power plants located in mining areas will generally have strong positive economic linkages with the mining sector and provide increased economic opportunities for the local population. Strong economic linkages to particular sectors in the local economy can add to the attractiveness of such projects. In most cases, by siting power plants in rural areas, the energy developer faces a situation where physical, economic, and political conditions are more conducive to the orderly and timely completion of the project.

The benefits that actually accrue from such developments to the local area have been a point of contention among energy impact and development researchers. R. L. Little and S. B. Lovejoy (1979) suggest that benefits are often dispersed over a much broader geographical area than that which is directly impacted by the construction of the power plant. During the construction phase of such projects, rural communities can experience a general social upheaval and local governments can be put under severe fiscal stress (Campbell 1976). The benefits that are realized by the development of electricity-generating facilities have been viewed by L. Susskind and M. O'Hare (1977) as only short term, temporarily stimulating growth in a very limited number of economic sectors.

In order to evaluate better the economic impacts of large-scale power-generating facilities, we apply quasi-experimental evaluation methods to four West Virginia counties where coal-fired power plants have been constructed.

Quasi-Experimentation and the Measurement of Impacts

When thinking about experiments and experimenting one generally conjures up a mental picture of a scientist working diligently in a laboratory, mixing chemicals or viewing some specific culture under a microscope. All experiments are a essentially a test, a test in which some phenomenon is being deliberately manipulated by the scientist. For the social scientist, tests are conducted outside the confines of the laboratory in the complex economic and social reality of the world around us.

In our case, the test is to determine how the introduction of large-scale coal-fired power plants into a rural area affects the economic development and growth of that area. Equally important to determine is how we can be sure that the changes that are observed can be causally linked to the introduction of the power plant. Answering these questions will shed new light on the benefits or costs communities gain from being host to such facilities as well as on the usefulness of such projects to a rural development strategy.

Experimental designs are generally made up of treatments (independent variables) and outcomes (dependent variables). In order to ensure that a particular treatment did cause a particular outcome, experimental designs make use of some kind of comparison or control-"without a control group there is no way to tell how much of the overall effect in the experimental group was true cause and how much was extraneous effect" (Bailey 1978). In social science research, this sorting out of real and extraneous effects has been accomplished by the assignment of random groups which are to receive or not to receive a specific treatment.

While a true experimental design affords the most control over nontreatment influences on outcomes, it suffers from a number of shortcomings that reduce its application to much social science research. Random assignment of treatment and nontreatment groups grows much more difficult as one moves out of the controlled confines of the laboratory. The researcher has little control over what area gets a power plant, a dam, or a major military base. In evaluating many social programs, the use of experimental designs would mean denying some eligible program participants aid or services.

The use of quasi-experimental designs in the examination of social phenomenon has become increasingly popular. Unlike true experimental designs, quasi-experimental designs do not utilize randomization of subjects before they enter designated experimental and control groups. In some quasi-experimental designs, control groups are not utilized. Replicated testing, removed treatment, and other innovations can approximate control groups in many settings. Such designs range in sophistication from the one group post-test only design to the more sophisticated time series designs. In most cases, these designs can be more readily adapted to social science problems. In trying to analyze the results from any experiment, the separation of the treatment effects and those caused by the initial noncompatibility in the treatment and nontreatment groups is critical (Cook and Campbell 1979).

For our purpose, we cannot randomly assign power plants to specific areas of the country. We can, however, identify areas (counties) that were similar to one another at time periods prior to one county's having a power plant built within it. It is then possible to observe how the impacted county economically changed compared to those that prior to the construction of the power plant were similar to it.

Since quasi-experiments lack randomization of subjects, the similarity between control and experimental groups is of the utmost importance. In evaluating economic changes in counties that power plant development took place in with otherwise similar counties, this book utilizes shift-share analysis and regional policy evaluation techniques developed by Barry Moore and John Rhodes (1973, 1974) and Andrew M. Isserman and John Merrifield (1982). These techniques identify possible control groups on the basis of their industrial structure.

Determining the Control Group

The shift-share method historically has been used to compare an area's economic growth performance to that experienced by the nation as a whole. The technique divides economic growth into three components: national growth, industrial mix, and regional share.

The national growth component represents the growth that would have occurred in a particular industry if its employment had grown at the same rate as the average for all industries combined.

Structural changes such as technological innovations and changes in demand patterns vary across industries. Some industries grow more rapidly than others. These relative employment changes make up the industrial mix component. Negative values indicate that employment grew at a slower rate in a particular industry than in all industries combined. A positive value indicates that employment in an industry grew at a faster rate than did employment for all industries combined. The sum of the national growth and industrial mix components for an area indicate the growth or decline in employment that would have taken place in the area if its industries had grown at the same rate as the national counterparts.

The regional share component consists of growth or decline in employment in an industry above or below the national rate of growth or decline for that industry. The sum of these three components comprise total employment change. Income can be substituted for employment in each component of shift-share analysis. Interpretation is then based upon comparative income changes as opposed to employment changes.

The regional share of shift-share analysis is critical in differentiating the impacts power plants have had on their host counties. Assuming that each local industry grows at the national growth rate, the economic impacts of power plant developments can be measured as the difference between an area's actual change and the expected change at national growth rates (Moore and Rhodes 1973, 1974).

The impact can be written algebraically as

$$I_s^+ = E_s^+ = -r_c^{o+} E_s^o \tag{6}$$

where I^+ represents the total impact in year + on the study area s, E_s^+ and E_s^o represent economic activity (total income in the study area in years + and 0, and r_c^{o+} the growth rate in income in the control area from year 0 to year +.

The right-hand side of the equation is the regional share in shift-share analysis. Power plant development is assumed to be entirely responsible for the differences between area and national industry specific growth rates. For this study, the industry specific growth rates of the control group replace the national growth rates in equation (6). The impacts of power plant development is then the difference in growth rates of the impact area and the control group counties.

The Impact Counties

The areas impacted by power plant development are Mason (Sporn, Mountaineer), Putnam (John Amos), Pleasants (Pleasants), and Grant (Mt. Storm) counties of West Virginia. Each of these counties is host for a coal-fired power plant. All of the plants have been constructed in the past twenty-five years. Each county is primarily rural in nature with few towns of over 2,500 persons in population. Also note the difference in the sizes of the power plants located in each county. The John Amos plant located in Putnam County is one of the world's largest coal-fired power plant. The Mountaineer plant located in Mason county is less than half its size.

Selecting Control Groups

Selecting appropriate control groups for each of the four West Virginia counties is a lengthy process. Initially, a calibration period and impact period have to be designated for each county. The calibration period is the time up to the beginning of the impact. The impact is defined simply as the year construction began on the power plant. In addition, some geographical region should be selected from which control group counties can be chosen.

In this study, the geographical region included the Southeast, Northeast, and New England, plus Ohio, Indiana, and Illinois for all the impact counties. Calibration periods begin from 5 to 7 years before the impact period. Calibration periods for each of the four West Virginia counties are as follows:

Mason (1970-1975)
Pleasants (1967-1973)
Putnam (1962-1968)
Grant (1959-1962).

Impact periods began in 1962 for Grant, in 1976 for Mason, in 1974 for Pleasants, and in 1969 for Putnam. Data availability reduced the length of the calibration period for Grant County.

The second step in this process involves drawing up a large list of potential control group counties. This is first done by the use of factor and cluster analysis, then by shift-share analysis. The most important selection criteria in this first cut are population size, industrial structure, and geographical location characteristics similar to the study county. Data from BEA income by sector tapes and county business patterns employment data were used. After this procedure, each of the impacted counties had between 55 and 100 potential control group county candidates.

Control group candidates were then evaluated on the basis of two data sets: the first consisting of percentage changes in one-digit sector Standard Industrial Classification codes for the calibration period, and the second consisting of more detailed indicators of a county's economic status prior to the impact period. Variables such as the relative sizes of one-digit industrial sectors, land area, per capita income, unemployment rate, education level, and percent of persons above sixty-five years of age and below eighteen years of age were included in the second data set. For each data set, factor then cluster analysis was used. Counties that were among the top twenty to twenty-five most similar to the study county in both cluster analyses made up a short list of control group candidates. Generally, a dozen counties made up the list for each of the impacted counties.

Finally, the measured policy/event impact should be zero during the calibration period. This confirms that the study area and the control group were developed similarly before the impact began. For each case, different combinations of control group counties were tested to find the smallest overall impact. These lowest impact combination counties formed the final control group for each study area. Appendix 1 indicates how close each group of control counties came to the impact county. As can be seen in each case, differences do exist from the ideal zero score. Isserman and Merrifield (1982) treated these differences as a quasi-experimental confidence level in their study on the impact of Economic Development Administration development grants. For example, in Mason County, a total of $370,000 in income was wrongly reported.

The Findings

The impact of power plant construction and operation on the four West Virginia counties is shown in Tables 27 to 30. In Mason County, most sector-specific results are about the size and sign one would have expected: large positive changes in transportation, communications, and public utilities (TCPU); construction, in the early part of the period; and in services: and large negatives in residential adjustment. Somewhat surprising were the large negative impacts on manufacturing and farming income and the lack of any significant impact on trade, especially retail trade. An attempt should be made to determine whether the slow growth (relative to the control group) of manufacturing and farming incomes are indeed linked to the construction and operation of the power plant. Total impacts were positive for each year of the impact period (1976-1984) through 1982. Income impacts peaked at $11 million in 1979. Impacts were negative thereafter. This suggests that income gains exceeded opportunity costs in the short run, but not in the long run.

In Pleasants County (Table 28), the study period results are again about what one would expect for most sectors, for most years. Construction and TCPU impacts are large and positive until near the end of the impact period. Through 1980, the residential adjustment impact is large and negative. Trade sectors witnessed zero or negative impacts: impacts on services were mostly small. Total impacts are in a positive direction for each year of the impact period. These range between $2 million and $27 million in added income in each year of the impact period.

Though they started more slowly than in the three other counties, the sectoral and temporal patterns of impacts in Grant County (Table 29) are again about what would be expected, with two exceptions. The first exception is the impact on mining and the really positive total impact on income in recent years. Much of Grant County's large coal-mining production is utilized by the Mt. Storm power plant. The location of many of the mines in the corner of the county could explain the county's large negative residential adjustment in most years of the study period.

The impact estimates for Putnam County (Table 30) are the most dubious of the four. The county's economy was influenced by several major exogenous shocks early in the study period (1968-1984). There was the shock of construction and operation of the power plant, plus construction of a section of interstate highway, and the closing of a major synthetic fibers plant. Impact estimates should be interpreted as the joint direct, indirect, and induced effect of all three exogenous shocks. Each of the events will no doubt dominate the others in particular years.

Construction, TCPU, and residential adjustment impacts are once again consistent with a priori expectations. Total impacts were large and positive throughout the impact period. Unlike in the other three counties, the trade sector and services sector impacts were mostly large and positive.

For the most part, the research findings verify the previous work done by Little and Lovejoy (1979) that the benefits of such projects are dispersed over a wider geographical area than the host counties. The substantial residential adjustment in each of the four cases makes it clear that many workers were commuting to the host county from nearby areas. This served to reduce drastically the income gains to the local population for most years. This

Table 27
Economic Impacts of Power Plant Construction and Operation in Mason County (1,000 $)

SECTOR	1976	1977	1978	1979	1980	1981	1982	1983	1984
FARMING AND AGRICULTURE	464	-276	-2473	-2661	-2070	-3716	-4167	-4086	-2528
MINING	-1731	-1883	-2478	-2720	-3001	-3012	-2817	-3644	-3486
CONSTRUCTION	9400	3595	35959	45319	23448	1205	-2073	-3599	-4600
MANUFACTURING	-1112	-2598	-4118	-3867	-4532	-6462	-9449	-14416	-16557
TRANSPORTATION, COMM. & PUBLIC UTILITIES	301	1848	10102	11452	15554	18189	1707	15437	17584
WHOLESALE TRADE	-645	-776	-1289	-942	-1172	-1422	-2520	-2336	-3211
RETAIL TRADE	338	-94	81	-73	39	-367	-789	-1201	-1556
FINANCE, INSURANCE, AND REAL ESTATE	45	-1	89	120	125	197	343	437	574
SERVICES	-495	2026	383	1514	2025	2202	2844	2534	2032
FEDERAL	20	288	254	56	109	-70	-211	-63	-381
STATE AND LOCAL GOVERNMENT	368	997	1333	2753	3126	3466	3909	3963	4431
RESIDENTIAL ADJUSTMENT	-2598	-14281	-32527	-40093	-31496	-8325	-404	-4540	-5470
TRANSFER PAYMENTS, DIVIDENDS, INTEREST, AND RENT	292	381	281	514	3750	3110	4094	4865	1446
TOTAL	4335	6412	5031	10945	4795	3618	6929	-6699	-12287

Table 28
Economic Impacts of Power Plant Construction and Operation in Pleasants County (1,000 $)

SECTOR	1974	1975	1976	1977	1978	1979	1980	1981	1982	1983	1984
FARMING AND AGRICULTURE	-31	-91	-92	-75	-94	-31	137	63	-99	-51	-18
MINING	1190	936	2878	2154	564	2908	3821	5586	6997	4930	4665
CONSTRUCTION	368	5479	10846	30254	58368	29605	21252	9091	3773	215	-2080
MANUFACTURING	2346	3978	5627	4853	5220	8570	5545	3299	1523	-4339	-2463
TRANSPORTATION, COMM. & PUBLIC UTILITIES	568	-402	342	1663	3148	5258	6770	8118	8385	11090	11712
WHOLESALE TRADE	-27	282	783	-79	-436	-417	-560	-597	-678	-658	-869
RETAIL TRADE	14	43	-19	-16	-105	-105	34	146	84	-84	56
FINANCE, INSURANCE, AND REAL ESTATE	-13	-84	-115	-115	-124	-255	-205	-165	-189	-166	-228
SERVICES	-25	228	374	213	215	204	672	178	1683	693	242
FEDERAL	19	-8	-17	28	17	-13	20	-85	-150	-162	-184
STATE AND LOCAL GOVERNMENT	-40	-66	345	632	773	323	-412	654	695	42	695
RESIDENTIAL ADJUSTMENT	-2083	-5909	-11613	-22869	-50362	-23244	-8390	1736	4288	6150	10867
TRANSFER PAYMENTS	47	-220	-321	-670	-1028	-1551	-1940	-2377	-2224	-1740	-2364
DIVIDENDS, INTEREST, & RENT	-265	-452	-392	-254	-186	441	74	407	294	628	575
TOTAL	2068	3714	8626	15719	15970	21693	26818	26054	24382	16548	20434

Table 29
Economic Impacts of Power Plant Construction and Operation in Grant County (1000 $)

SECTOR	1965	1966	1967	1968	1969	1970	1971	1972	1973	1974	1975	1976	1977	1978	1979	1980	1981	1982	1983	1984
FARMING AND AGRICULTURAL	-167	-186	-13	-235	-22	585	511	579	587	1154	303	697	596	78	247	798	677	-1158	-1429	-1618
MINING	3163	4181	5233	4804	5540	6187	5733	6020	7678	9615	13163	4915	15297	16306	25624	32430	26167	34017	34158	42939
CONSTRUCTION	9	6	18	429	720	2530	9946	17638	13699	-484	-216	165	1345	2902	1736	1607	4531	4674	9781	6266
MANUFACTURING	-139	-124	-37	-183	-617	-588	-557	-1467	-1141	-267	302	-1670	-2698	-4209	-2664	-2882	-3760	-3087	-4591	-5675
TCPU	810	806	1300	986	1221	1458	2276	2017	2238	3426	4003	4954	5577	3633	3525	7236	7592	10510	17011	17656
WHOLESALE TRADE	324	476	537	-2911	-3399	-3738	-5182	-6245	-7686	-9014	-10154	-10395	-11198	-12948	-16301	-16747	-18232	-15952	-14814	-18383
RETAIL TRADE	-6	-19	-9	-12	-92	-237	-143	69	26	-93	-111	-176	-389	-651	-719	-437	-417	-476	-373	-386
FIRE, INSURANCE, AND, REAL ESTATE	-27	-39	-51	-67	-82	-136	-251	-354	-268	-245	-213	-315	-349	-417	-436	-549	-520	-567	-498	-360
SERVICES	209	145	-197	-398	-288	53	54	152	265	-325	-226	-863	-979	224	917	-344	2503	4945	3100	3070
FEDERAL	-1	19	-99	8	27	57	78	7	76	64	173	205	278	261	278	203	279	183	156	247
STATE AND LOCAL GOVERNMENT	105	151	191	221	210	274	296	399	532	596	740	997	1341	1356	1836	3398	3098	3117	3080	3098
RESIDENTIAL ADJUSTMENT	-2844	-3858	-4933	-3761	-5590	-8394	-15161	-19702	-17795	-7831	-9339	-10206	-11056	-11948	-18987	-25398	-25860	-34354	-40488	-43387
TRANSFER PAYMENTS	652	657	244	-25	94	186	238	366	509	742	412	691	850	316	-174	-179	-195	-374	661	578
DIVIDENDS, INTEREST, AND RENT	-39	58	113	84	51	25	61	244	306	413	186	434	559	1028	1294	1548	1755	2006	2033	2608
TOTAL	2049	2273	2297	-1060	-2227	-1738	-2101	-277	-974	-2249	-997	-567	-826	-4069	-3824	684	-2382	3484	8087	6653

Table 30
Economic Impacts of Power Plant Construction and Operation in Putnam County (1,000 $)

SECTOR	1969	1970	1971	1972	1973	1974	1975	1976	1977	1978	1979	1980	1981	1982	1983	1984
FARMING AND AGRICULTURE	460	529	371	455	212	608	313	460	446	-58	97	814	1144	682	331	-1423
MINING	9	-46	17	-11	-152	688	219	615	257	-450	-1	327	1451	854	314	687
CONSTRUCTION	7206	20610	20778	13557	9790	5692	7976	11599	15157	7982	4331	637	-2928	-2121	-2509	-8199
MANUFACTURING	-1327	-656	-958	-311	-74	629	-685	-4864	-3782	-19717	-22341	-20104	-12845	-4858	-9308	-13394
TCPU	678	502	2217	2348	3805	6884	6875	10294	11897	13891	19455	18451	20646	21427	21041	23664
WHOLESALE TRADE	-161	283	1056	1405	3571	4072	6669	6033	6725	7674	8827	11414	14715	16942	16504	15840
RETAIL TRADE	-155	323	20	274	1084	1525	2501	2360	2730	3282	3921	9106	10800	13226	15982	21503
FINANCE, INSURANCE, AND REAL ESTATE	13	-63	-57	82	-281	25	372	734	925	1413	736	613	231	340	886	1112
SERVICES	-95	-3	4228	8649	9034	2875	280	857	768	342	-47	47	3720	3628	3887	7859
FEDERAL	-84	-30	-72	211	232	260	314	362	802	703	877	990	780	390	355	301
STATE AND LOCAL GOVERNMENT	-115	-77	-71	61	401	668	1018	1823	3094	3473	3974	4140	5517	6212	5181	5260
RESIDENTIAL ADJUSTMENT	-8139	-16630	-20117	-20900	-20663	-11897	-5544	-7341	-10532	-1071	8611	5492	-20042	-27472	-36935	-53835
TRANSFER PAYMENTS	251	-55	-413	-859	-1002	-777	688	957	930	1115	1250	1121	1478	594	-114	-1423
DIVIDENDS, INTERESTS, AND RENTS	243	439	590	1147	1533	1650	1436	981	1665	2388	3357	8911	8557	7386	8873	9832
TOTAL	-1216	5126	7589	6108	7490	12902	22432	24870	31082	20967	33047	42049	33224	37230	24488	7784

was especially true during the construction phase of each project. In Grant County, where the Mt. Storm plant approximates what is a mine-mouth facility, it seems apparent that a substantial part of the mining workforce comes from outside the county.

Economic linkages with other sectors outside of construction and TCPU are not self-evident. Clear patterns across all four counties are not present. This is comparable to Susskind and O'Hare's findings (1979) that such developments only impact a select number of sectors. However, while Susskind and O'Hare suggest that the positive impacts of such developments are short term, our findings show that impacts can be positive for substantial periods of time. In the Pleasants and Putnam county cases, impacts have remained positive for a decade or longer.

The quasi-experimental evaluation method used here is in the early stages of development. In their study of the income impacts of EDA development grants and energy developments in western boomtowns, Isserman and Merrifield (1982, 1985) indicate that this method does not rule out all the earlier mentioned threats to causality. We cannot control for other events that have occurred in either the impact counties or the control groups. As noted earlier, the control group counties are not perfect in respect to the affected areas. Even with these limitations, however, the findings are significant in that they substantiate the results of past research.

As a development strategy for local areas, power-generating facilities are neither a cure nor a failure. For the host county, economic leakages from commuting workers greatly reduce income gains to the local population. However, the plants do tend to provide positive income gains for some areas over extended periods of time. The scope of this section focuses only on countywide impacts. Power plant development should probably be viewed in the context of larger geographical areas in order truly to assess their value as a means for rural development.

ELECTRICITY AND INDUSTRIAL LOCATION

In addition to the jobs and revenues gained by having coal-fired power plants in a region, the availability of a long-term source of reliable and relatively inexpensive electricity may also aid in attracting certain firms to a region. Regions with a dominance of coal-fired power plants tend to have lower than average electricity prices. Table 31 illustrates the variation in industrial electricity prices in the United States. The New England states are well above the national average, and the midwestern states are closer to the national average. West Virginia and other Appalachian states have lower electricity prices; prices are the lowest in the West, except California.

Many utilities promote their abundant and inexpensive electricity supplies in order to attract new firms into their service district. These promotional materials assume that firms view electricity as being one of the more important reasons in selecting a site for a facility. Some studies support this view, but others are not so supportive

Table 31
State Average Industrial Electricity Prices (dollars per million Btu)

	1978	1979	1980	1981	1982	1983	1984	1985
AL	7.61	8.92	10.43	12.04	13.74	14.61	14.82	14.98
AK	10.17	10.23	10.85	9.36	22.50	18.83	21.06	20.32
AZ	9.10	10.09	11.54	11.95	16.95	15.48	15.82	15.67
AR	7.52	7.88	9.13	10.33	11.70	12.07	12.95	14.41
CA	10.83	11.02	15.76	17.75	21.18	19.89	20.42	22.81
CO	6.95	8.05	9.67	10.23	12.53	12.35	12.98	13.24
CT	10.92	12.93	16.46	20.81	21.73	22.06	23.39	22.84
DE	10.67	11.66	14.72	16.35	17.72	15.20	16.07	16.36
FL	9.16	10.17	13.64	16.88	16.61	16.50	17.37	17.41
GA	9.03	9.46	10.69	11.26	13.55	13.31	13.92	13.60
HI	13.19	14.26	18.87	30.51	31.76	27.73	28.08	26.11
ID	3.83	4.46	5.52	6.33	6.61	7.26	7.50	9.01
IL	9.05	9.74	11.99	13.38	15.34	17.25	16.13	16.00
IN	9.18	9.77	11.00	12.07	13.95	14.02	14.98	15.13
IA	8.37	9.41	10.34	11.64	12.84	13.46	14.31	14.02
KS	8.64	9.10	10.56	11.98	14.32	15.70	15.82	15.31
KY	6.84	7.59	8.69	11.01	11.27	12.25	13.19	14.68
LA	5.58	6.79	9.07	11.48	13.12	13.59	13.22	14.14
ME	6.99	9.09	13.26	14.91	14.29	13.77	15.19	15.79
MD	10.14	10.01	11.80	12.92	14.57	13.74	13.64	14.50
MA	12.73	14.15	17.60	22.15	20.68	21.59	22.96	21.10
MI	10.36	11.11	13.32	14.85	16.48	16.19	16.64	17.50
MN	9.24	9.58	11.07	12.23	14.17	13.74	13.16	13.18
MS	8.76	9.92	11.48	14.01	15.37	14.46	14.25	14.53
MO	8.97	9.69	11.16	11.76	13.09	13.28	12.86	13.60
MT	2.28	2.50	3.01	3.93	6.42	7.35	7.78	7.65
NE	6.59	7.11	8.06	9.45	11.11	11.73	11.44	11.55
NV	7.69	8.97	11.77	12.04	13.37	14.74	13.07	13.45
NH	11.49	12.62	15.98	21.75	20.13	20.66	21.57	20.14
NJ	12.22	13.21	17.19	19.97	22.22	21.72	22.48	23.47
NM	9.97	12.32	13.64	15.19	17.35	18.02	16.73	16.66
NY	9.32	10.48	12.24	15.38	16.39	15.70	17.12	16.00
NC	8.01	8.24	9.04	10.83	11.88	12.28	14.22	15.55
ND	9.28	9.43	9.89	11.92	14.17	15.32	15.79	15.94
OH	7.57	8.26	9.89	10.83	12.93	13.15	12.43	12.21
OK	7.32	7.60	8.85	10.83	12.01	12.87	13.43	14.05
OR	3.28	3.75	4.69	5.93	8.52	9.74	10.20	10.77
PA	9.75	10.49	13.00	14.48	16.95	16.22	16.88	17.77
RI	13.01	14.67	18.65	23.18	22.50	22.96	24.39	22.84
SC	7.25	8.00	8.63	9.92	11.39	11.60	11.98	12.52
SD	8.03	8.33	9.29	10.70	12.62	12.81	13.04	12.91
TN	7.41	8.68	9.83	11.96	13.67	13.74	14.07	14.80
TX	7.43	8.56	9.99	12.39	14.88	15.36	14.86	14.71
UT	7.58	8.72	10.05	11.45	12.90	13.53	13.92	15.04
VT	8.88	9.00	10.53	14.16	16.02	15.67	17.40	19.12
VA	8.44	10.09	12.24	12.23	13.83	13.84	13.31	12.97
WA	1.45	1.57	2.25	3.03	5.59	5.71	7.17	6.45
WV	7.64	7.63	8.69	9.64	11.48	11.57	11.01	11.22
WI	8.14	8.93	10.05	11.39	13.27	13.31	13.19	12.97
WY	4.86	4.88	5.20	6.33	8.98	10.18	10.38	10.53

Source: U.S. Energy Information Administration, State Energy Data Book, 1987.

of this claim (Calzonetti and Walker 1988). It is clear, however, that electricity prices and reliability do play an important role in some industries.

Tables 32, 33, and 34 provide generalized cost breakdowns for manufacturing plants currently operating in West Virginia. Plant managers were contacted and asked to provide information of the plant specifications for non-tax related locally varying costs. Plants include a fabricated plastics product plant with 150 employees, a women's clothing plant with 169 employees, and an industrial inorganic chemical plant with 800 employees. Comparison states have also been provided to give an indication of the range of costs for each of these factors.

Using state electricity price averages provided by the U.S. Energy Information Administration, one can compute the cost savings available by locating in a state with high versus low electricity prices. For the industrial inorganic chemical plant employing 800 workers, for instance, the total labor costs will be $26,601,803 in West Virginia; $28,417,386 in Texas; and $25,148,631 in New Jersey. Electricity costs will be $61,416,900 in West Virginia; $77,128,200 in Texas; and $95,696,100 in New Jersey. Selecting the cheapest labor cost state (New Jersey) saves $3.2 million annually, but $34.3 million is saved annually by selecting the state with the cheapest electricity rates. This example is extreme, but it does indicate the importance of electricity prices to select industrial plants.

A survey research study was conducted in West Virginia and eastern Kentucky to identify the importance of electricity, and other factors, in the search for new manufacturing plants (see Calzonetti and Walker 1987). A total of 162 plants that have located in the region since 1978 still survive. Seventy of these are in West Virginia; ninety-two are in eastern Kentucky. Thirty-eight of the eastern Kentucky plants participated in the study, and thirty-six of the West Virginia plants participated, providing a response rate of 45.6 percent. Plants were distinguished between those that are electricity intensive and those that are not electricity intensive, and the results to questions regarding the significance of electricity were evaluated.

Table 35 compares the responses to a question asking whether electricity prices were important in the choice of the plant location at two levels in the search process, the regional level and the local level. Only twenty-six of seventy-four plants record having undertaken a regional level search. Of those conducting a regional search, precisely half considered electricity to be an important factor. More of the plants that were not electricity intensive said that electricity was important than those that were electricity intensive.

Many more of the plants reported undertaking a local level search. Fifty of the seventy-four plants were located after completing a local search. Only fourteen of these fifty plants reported that electricity was important in their search. About the same percentage of the electricity-intensive plants reported electricity to be important as those for whom electricity is not considered a

Table 32
Comparative Cost Analysis for Fabricated Plastics Products (SIC 307), 1986.

Cost Factors	Quantity[b]	West Virginia	Kentucky	Ohio	Pennsylvania
Labor					
Production	120[c]	$3,493,893.3	$3,619,534.5	$3,138,255.6	$2,622,424.9
Clerical	20[d]	$ 352,560.0	$ 330,720.0	$ 429,520.0	$ 388,960.0
Management	10[e]	$ 346,092.5	$ 333,063.0	$ 353,901.0	$ 352,337.0
TOTAL LABOR	150	$4,192,515.8	$4,283,317.5	$3,921,776.0	$3,363,721.9
Energy and Water					
Electricity (m kWh/yr.)	3.6	$ 160,218.0	$ 141,588.0	$ 197,478.0	$ 212,382.0
Natural Gas (mcf/yr.)	6,218	$ 28,574.1	$ 28,059.2	$ 31,534.5	$ 29,475.1
Water[a] (m gal./yr.)	77.7	$ 154,003.2	$ 76,590.0	$ 43,968.5	$ 76,639.5
Sewage[a] (included with water)	--	$ 49,951.2	$ 65,627.8	$ 65,902.9	$ 58,666.4
Depreciation					
Land (acres)	11.5	$ 305,037.5	$ 366,045.0	$ 170,821.0	$ 427,052.0
Building (sq ft)	100,000	$ 96,222.2	$ 89,622.2	$ 97,155.5	$ 97,155.5
Transportation	--	$ 173,475.0	$ 173,475.0	$ 173,475.0	$ 173,475.0
TOTAL COSTS (1986)		$5,160,027.0	$5,224,324.7	$4,702,112.0	$4,438,567.4

Notes: [a]Water and sewage rates calculated from data for Charleston, W.Va., Louisville, Ky., Cincinnati, Ohio, and Philadelphia, Pa.
[b]Adopted from Fantus Corporation.
[c]Bureau of Labor Statistics, 1985.
[d]Administrative Management Society, 1985.
[e]Administrative Management Society, 1982.

Source: Data collected by authors.

large input. In sum, the results of this survey suggest that very few manufacturers consider electricity in their search for new plants.

It must be remembered that electricity is important to a minority of manufacturing plants and should not be dismissed as an industrial location factor. Electricity

Table 33
Comparative Cost Analysis for Industrial Inorganic Chemicals (SIC 281), 1986.

Cost Factors	Quantity	West Virginia	Texas	New Jersey
Production Workers	600	$20,549,448	$21,962,700	$18,856,092
Clerical Workers	50	$ 847,500	$ 1,022,500	$ 1,052,500
Managerial Workers	150	$ 5,204,925	$ 5,432,186	$ 5,240,039
Total Labor	800	$26,601,873	$28,417,386	$25,148,631
Electricity	115,000,000 kWh/mo.	$61,416,900	$77,128,200	$95,696,100

Source: Data collected by R. Walker and F. Calzonetti.

prices and availability are important to some firms, and firms in these industries should be targeted in promoting regional development. Also, other attributes of the Appalachian region may reduce its overall attractiveness, and it may be difficult to transfer the findings of this study to other regions.

PROMOTING ELECTRICITY

As noted earlier, many utilities have generating capacity in excess of their reserve margin needs and are promoting efforts to increase the use of their existing base load generating capacity. Most utilities are facing a short-term excess supply situation, and these efforts are aimed at ways to remedy the situation over the next five or so years. Other utilities wish to enter agreements to sell electricity for longer periods of time, and they are considering from twenty- to thirty-year firm power contracts. In addition, some areas, in concert with the electricity industry in some cases, are promoting electricity export sales as a long-term strategy. For the electric utility, electricity sales represent a source of revenue to the seller. The state, province, or locality may view electricity as a means to provide jobs and revenue over a long period.

Area Development Rates

In recent years, many utilities in the United States' manufacturing belt have been offering area development rates or promotional rates designed to attract industry into regions with stagnant economies. With the industrial load being the largest and most reliable user of electricity for most utilities, many electricity companies were hard hit by industrial plant closures or plant reductions caused by the recent recession. The industrial load is usually consistent and often very large.

Table 34
Comparative Cost Analysis for Women's Clothing (SIC 233), 1986.

Cost Factors	Quantity[b]	West Virginia	Massachusetts	New York	Pennsylvania
Labor					
Production	165[c]	$2,184,811.2	$2,461,413.9	$2,829,050.4	$2,282,847.6
Clerical	3[d]	$ 50,850.0	$ 54,600.0	$ 55,200.0	$ 56,100.0
Management	1[e]	$ 34,699.5	$ 35,891.7	$ 33,449.7	$ 35,325.6
TOTAL LABOR	169	$2,270,360.7	$2,551,905.6	$2,917,700.1	$2,374,273.2
Energy and Water					
Electricity (kWh/yr.)	372,631	$ 16,583.9	$ 18,049.4	$ 29,735.3	$ 22,330.4
Natural Gas (mcf/mo.)	157	$ 720.6	$ 654.6	$ 847.8	$ 744.1
Water[a] (gal./mo.)	43,300	$ 971.6	$ 587.1	$ 270.1	$ 415.6
Sewage[a] (included with water)	--	--	--	--	--
Depreciation					
Land (acres)	5	$ 198,937.5	$ 119,362.5	$ 79,575.0	$ 185,675.0
Building (sq ft)	24,000	$ 23,093.3	$ 23,093.3	$ 23,770.7	$ 23,317.3
TOTAL COSTS (1986) (nontransportation)		$2,510,667.6	$2,713,652.5	$3,051,899.0	$2,606,755.6

Notes: [a]Water and sewage rates calculated from data for Charleston, W.Va., Boston, Mass., Rochester, Ny., and Philadelphia, Pa.
[b]adopted from Fantus Corporation.
[c]Bureau of Labor Statistics, 1985.
[d]Administrative Management Society, 1985.
[e]Administrative Management Society, 1982.

Source: Data collected by R. Walker and F. Calzonetti.

Economic development or promotional rates lower the price of electricity to an industrial customer for a specified period of time to encourage the plant to come into the region. The electric utility gains directly from the new plant's demand and indirectly from the workers attracted with the plant and other multiplier effects. It is not clear if these rates indeed attract new industry, but electricity prices are not the dominant factor in selecting manufacturing locations for many industries.

Table 35
Importance of Electricity Prices: Regional and Local Search
(New West Virginia and Eastern Kentucky Plants, 1978-1986)

	Importance of Electricity Prices at the Regional Search Level: (compared regions in search for plant site)		
	Important	Unimportant	Did Not Search
Electricity-Intensive	5	6	11
Non Electricity-Intensive	8	7	37
	Importance of Electricity Prices at the Local Search Level: (compared local sites in search for plant site)		
	Important	Unimportant	Did Not Search
Electricity-Intensive	5	12	5
Non Electricity-Intensive	9	24	19

Source: Calzonetti and Walker, "The Regional and Local Search for New Manufacturing Plants," 1986.

Utilities recognize that industrial customers are an important component of their load, and many seek immaginative ways to keep plants in the service district profitable. Some utilities offer energy audits designed to reduce electricity costs, and some manufacturers can take advantage of off-peak rates to reduce their bills. There are even agreements that link electricity rates to the price of the manufacturer's product. Ohio Power Company has an agreement with Kaiser Aluminum that links the charge for electricity to the worldwide price of aluminum ingots. Electricity constitutes approximately one-third of the total cost of operating Kaiser's primary aluminum reduction plant, and the power company would prefer to keep the plant operating, albeit at lower capacity, at times when aluminum prices are lower than to have a total plant shutdown (Public Utilities Commission of Ohio 1985).

Electricity Exports

Utilities with excess generating capacity look for buyers in their own region to make the best use of their facilities. An official of Consolidated Edison of Chicago, for instance, said that his company sent letters of inquiry to utilities adjacent to their system to see if they would be interested in purchasing power not needed internally in the short term. When few responded with interest, the Chicago utility sent letters to utilities not adjacent to Con Edison. In the East, utilities with nuclear facilities prefer to keep them operating continuously for economic and engineering reasons. At night, when local demand is low, utilities sometimes give power away to other utilities who are able to use the power. It is expected that this practice will continue only a short time until the demand for electricity rises.

It is the long-term power trade that is of interest in this book. Long-term power trade will involve firm power sales for a twenty- to thirty-year period meaning that a portion of a utility's capacity will be made available to serve the export market. Few electric utilities actually undertake such long-term agreements, but it appears that the view of such transactions is growing more favorable. Utilities that have been exporting power in recent years resulting from temporary supply-demand imbalances and regional differences in electricity-generating costs grew to appreciate the significant revenues from such transactions.

State governments have also recognized the important economic benefits of power trade. Illinois, Indiana, Ohio and West Virginia have recognized the importance of power trade to their state economy. In Indiana, Hoosier Energy has been exporting about 400 megawatts of electricity from its Merom plant to Virginia Power since January 1985. The annual revenues gained from this sale has resulted in an estimated 43 percent savings in projected electricity price increases for the company's customers.

West Virginia is the state most actively trying to promote electricity exports as a state economic development strategy. Electricity exports provide jobs in the related coal-mining industry and substantial tax revenues to state and local governments. Electricity production has grown rapidly in the state over the past fifteen years. In 1970, West Virginia produced 33,000 gigawatt-hours of electricity. Electricity exports are generally viewed favorably in West Virginia, even among the leading state environmental leaders. Electricity exports can be increased by increasing the utilization the of existing generating capacity or by constructing new power plants.

CONCLUSION

Using West Virginia as an example, this chapter has shown that the electricity industry and electricity-generating facilities provide important economic benefits to local and state economies. Benefits also result when power is exported from the state to consumers elsewhere. These impacts depend upon the level of usage of these facilities and the origin of the coal, and they vary between the construction and operation period. Also, electricity is an important factor in certain manufacturing plants, but it generally is not recognized as a significant location factor for manufacturers seeking new plant locations. Given the economic gains by increasing power sales to other regions, there is great interest in developing more permanent power trade arrangements.

BIBLIOGRAPHY

Bachtel, D. C. and J. J. Molnar. 1981. "Women As Community Decision Makers." Human Services in the Rural Environment 6, 2:3-10.
Bailey, Kenneth O. 1978. Methods of Social Research. New York: Free Press.
Calzonetti, F. J. and R. T. Walker. April 22, 1988. "The Regional and Local Search for New Manufacturing Plants: Do Electricity Prices Matter?" Paper presented at the Industry Location and Public Policy Symposium, Knoxville, Tenn.
Campbell, K. A. 1976. Nuclear Power Plant Development: Boom or Bust? Washington, D.C.: National Association of Counties.
Cook, Thomas and D. J. Campbell. 1979. Experimentation. Chicago: Rand McNally.
Greenstreet, David. 1987. A Users Guide to the 1982 West Virginia Input-Output Model. Morgantown, W. Va.: Center for Economic Research, West Virginia University.
Isserman, Andrew M., and John Merrifield. 1982. "The Use of Control Groups in Evaluating Regional Economic Policy." Regional Science and Urban Economics 12 43-58.
Isserman, A. M. and J. Merrifield. 1985. "Quasi-experimental control group methods for regional analysis: An Application to an Energy Boomtown and Growth Pole Theory" Research paper 8610. Morgantown, W.Va.: Regional Research Institute (Nov.).
Little, R. L., and S. B. Lovejoy. 1979. "Energy Development and Local Employment." The Social Science Journal 16 (April):27-49.
Miernyk, William H. 1965. The Elements of Input/Output Analysis. New York: Random House.
Moore, Barry, and John Rhodes. 1974. "Regional Policy and the Scottish Economy." Scottish Journal of Political Economy 21:215-235.
---. 1973. "Evaluating the Effect of British Regional Economic Policy." Economic Journal 83:87-110.
Murdock, S. H., and F. L. Leistritz. 1979. Energy Development in the Western United States: Impacts on Rural Areas. New York: Praeger.
Public Utilities Commission of Ohio. 1985. Power Agreement between Kaiser Aluminum and Chemical Corporation and Ohio Power Company. Columbus, Ohio: Public Utilities Commission of Ohio.
Stout-Wiegand, N., and R. B. Trent. 1983. "Sex Differences in Attitudes toward New Energy Resource Developments." Rural Sociology 48, 4:637-46.
Susskind, L., and M. O'Hare. 1977. Managing the Social and Economic Impacts of Energy Development. Cambridge, Mass.: Massachusetts Institute of Technology, Laboratory of Architecture and Planning.
U.S. Energy Information Administration. 1987. State Energy Data Book. Washington, D.C.: U.S. Energy Information Administration.
West Virginia Coal Association. 1986. Coal Facts 1985. Charleston, W. Va.: West Virginia Coal Association.

7

The Politics of Acid Rain and Canadian Power Imports

Gregory G. Sayre and Frank J. Calzonetti

The topics of acid rain control and Canadian power imports weigh heavily in discussions of coal-by-wire. Acid rain legislation, if passed, will have an effect on the competitiveness of most existing coal-fired generating facilities in the Appalachians and Midwest, and it could result in some substitution of low-sulfur coal for high-sulfur coal. Also, if government policymakers and utility officials view Canadian power imports to be an acceptable partial solution to energy and acid rain problems then the contribution of coal-by-wire will surely diminish since, under most circumstances, coal-by-wire cannot compete with Canadian hydroelectric power.

We discuss these issues together in this chapter because they are so inseparable in many respects. On one hand, those groups promoting the increased use of Canadian hydroelectric power point to acid rain as a problem resulting from the use of coal in many power plants. On the other hand, some coal interests view Canada's strong stand against acid rain as a strategy to reduce the competitiveness of coal. In this chapter, the nature of acid rain is discussed and then the components of proposed legislation and implications for coal-by-wire. Consideration is then given to the Canadian question as it pertains to congressional legislation and coal-by-wire.

THE ACID RAIN ISSUE

The 1940s and 1950s witnessed major Appalachian industrial cities such as Birmingham and Pittsburgh struggle with life-threatening air quality problems. The emissions from coking plants, power plants, steel mills, and steam engines sometimes became trapped in river valleys under stagnant atmospheric conditions resulting in dangerously high levels of atmospheric pollutants. In Donora, Pennsylvania, for instance, twenty people died in 1984 because a low-level inversion trapped pollutants in the

Monongahela River valley (Strahler and Strahler 1983, 111).

The Clean Air Act and its amendments as well as changing economic conditions basically eliminated the local air-pollution problem. When steam engines were replaced by diesel engines in the 1950s and when steel mills and coking ovens closed during the 1960s and 1970s, the air in many Appalachian cities improved considerably. Power plants were required to install electrostatic precipitators to control the emissions of particulates (which used to fall in the immediate area downwind from the stacks), and in the late 1970s new power plants were required to install scrubbers to reduce the emissions of sulfur dioxide (SO_2). Also, many utilities installed tall stacks to disperse pollutants to meet ambient air quality standards.

The Clean Air Act and its amendments failed to address a problem that was not as obvious as the air quality problems previously experienced: the long-range transport of pollutants that were invisible to the naked eye. It is difficult to discern any emissions when viewing a power plant stack that is equipped with an electrostatic precipitator. However, large quantities of SO_2 and nitrogen oxides (NO_x) are being emitted and are usually carried aloft and away from the area. The impacts of this pollution are experienced some distance from the source. It has been difficult to link events in one region to a pollution source in another region. Also, it has proved to be almost impossible to arrive at a politically acceptable solution to the problem without harming one region's interests.

Those who perceive acid deposition to be a major environmental problem generally maintain that (1) acid deposition is for the most part attributable to man-made emissions of sulfur and nitrogen; (2) these emissions are transformed in the atmosphere to sulfuric and nitric acids and can be transported long distances; and (3) harm results to aquatic ecosystems, forests, crops, materials and buildings, and humans and there is a general decline in visibility. On the other side, there are those who maintain that (1) acidification is more likely a result of physical processes or cycles that are not very well understood and that man-made origins are overstated; (2) it is not yet clear whether acidification is increasing; (3) local pollution sources are more important in the acidification process than long-range transported pollutants; and (4) the complexity of atmospheric transformation of pollutants makes it difficult if not impossible to understand the relationship between sources and areas affected by deposition.

The Physical Problem

The term "acid deposition" should be used rather then the term "acid rain" because the process involves the deposition of a variety of pollutants in both wet and dry forms. The primary pollutants involved include SO_2, NO_x, and hydrocarbons which are transformed in the atmosphere into the secondary pollutants of ozone, sulfate, and nitrate. These secondary pollutants contribute to haze, mainly by airborne sulfates (thus impairing visibility), and fall to the ground as direct deposition, or as acid precipitation in conjunction with rain, snow, fog, or dew.

Acid deposition has been implicated in lake acidification, materials damage, adverse effects on human health, damage to forests and other vegetation, and damage to other ecosystems. The damage from acid deposition depends upon the quantity received and the sensitivity of the receiving area to the pollutants. The sensitivity of the area depends upon the season, geology, vegetation, and degree of natural stress the environment is under.

Damage to aquatic ecosystems has been most clearly documented. Acid deposition, either falling directly on lakes or on watersheds, reduces the lake's pH, which cannot be tolerated by many species of fish. In addition, acids aid in releasing toxic metals, particularly aluminum, which are carried into aquatic ecosystems poisoning fish and other organisms. Toxic metals, including lead and mercury, can also be released and enter the water supplies used for human consumption in these regions. According to the Ontario Ministry of the Environment, about 20 percent of the lakes in Eastern Canada are acid altered (Office of Technology Assessment 1985, 43). The New York State Department of Environmental Conservation reported that fish populations are endangered in over half of all the lakes and ponds in the Adirondack region (Canadian Embassy 1984, 2). The sensitivity of certain watersheds is increased because of the lack of buffering rocks and the natural acidity of the soil. Some watersheds are protected somewhat because of the underlying limestone or dolomite rock which maintains higher pH levels.

There has also been reported damage to forests in the Adirondack Mountains and the White Mountains, and in other forests in New Jersey, Tennessee, and Virginia. In many cases, these forests are at higher elevations and are under great natural stress from wind and temperature, but it is being suggested that acid deposition and ozone are contributing to the destruction of forests. Acid deposition can damage trees directly by removing essential nutrients from foliage, or indirectly by altering forest soils (Office of Technology Assessment 1985, 44). It is suggested that ozone and acid deposition also cause crop damage.

Other impacts include damage to structures, visibility reductions, and possible health effects. Air pollution, in general, has been shown to cause damage to building materials, but it has been difficult to assess the contribution that acid deposition makes to this problem. Health effects are also difficult to evaluate, although the smaller airborne particles, such as sulfates, are small enough to penetrate deeply into the lungs which could lead to increased mortality.

Wet acid deposition is measured by collecting stations in the United States which show that the most acidic precipitation occurs in eastern Ohio, western Pennsylvania, and northern West Virginia. Wheeling, West Virginia, recorded the most acidic rainfall in the fall of 1978 with a pH less than 2, and the western ridges of the Appalachians often record some of the most acidic precipitation in the country. While an exact dollar amount of the damage done by acid precipitation to Appalachia is not known, Appalachia does receive substantial quantities from both in-state and out-of-state sources, as indicated in Table 36. In a recent public opinion poll conducted by West Virginia University (Hunter et al. 1988), 80 percent of the West Virginia respondents believed that acid rain was at least a minor problem in the state. Surrounding this region and extending from the southern Appalachians to the Maritime Provinces is an area of less severe but clearly acidic precipitation.

Origins

Sulfur and nitrogen oxides originate from both man-made and natural sources; the man-made sources by far override the natural causes in the production of these gases. Natural sources of these gases include the ocean, volcanoes, lightning, wildlife, decomposing organic matter, and plant life. The Mount St. Helens volcanic eruption was the greatest single natural source of SO_2 in North America and it released at most 300,000 tons of SO_2 in a one-year period. According to the Office of Technology Assessment (1985), about 27 million tons of SO_2 and 21 million tons of NO_x were emitted in the United States in the year 1980 (Office of Technology Assessment 1985, 5). Over 90 percent had utility or industrial origins. Moreover, approximately 80 percent of the SO_2 and 65 percent of the NO_x originated from the states either bordering or to the east of the Mississippi River. Table 37 shows the total emissions of SO_2, NO_x, and hydrocarbons by state. The top five emitters of SO_2 were Ohio, Pennsylvania, Indiana, Illinois, and Missouri. It is estimated that 65 percent of the man-made SO_2 produced in the United States originates at electrical generating stations, and about 30 percent of the NO_x comes from electric utilities.

THE CLEAN AIR ACT

The Clean Air Act of 1970 is the major piece of federal legislation that deals with the regulation of air pollutants. The act has two major components to control air pollution and to ensure public safety. First, it provided for the establishment of the National Air Ambient Quality Standards by the U.S. Environmental Protection Agency; and second, it set emission limits on new pollution sources through the New Source Performance Standards (NSPS). The act as it now stands may not be designed to deal effectively with the problem of acid rain; in fact, it may have even been a contributing cause of the problem.

Air Ambient Quality Standards are made up of primary standards and secondary standards. Primary standards are set to ensure the public health and maintain an adequate margin of safety. Secondary standards are designed to protect public welfare, which includes pollution effects on soils, water, crops, vegetation, wildlife, and climate (Parker and Blodgett 1985). The act sets specific deadlines for the attainment of primary standards but not

Table 36
Interstate Emissions Transport (1000 tons SO_2).

	Own Emission Received	Emission Received from Midwest	Emission Received from South
Alabama	92	42	236
Arkansas	20	48	52
Florida	146	8	188
Georgia	158	60	392
Kentucky	108	274	226
Louisiana	42	10	68
Maryland	16	62	56
Mississippi	30	30	118
North Carolina	90	106	296
South Carolina	40	32	116
Tennessee	116	128	272
Virginia	30	188	224
West Virginia	62	236	140

Source: National Clean Air Coalition, 1981, 9.

secondary ones. Most of the effects of acid rain come under secondary standards. There are also provisions for the Prevention of Significant Deterioration (PSD) to prevent pollution sources from diminishing the quality of air in certain pristine regions.

The Clean Air Act was initially concerned with local pollution problems and concentrations above primary standards. As a way to reduce local concentrations many polluting facilities diluted local concentrations by raising the height of their smokestacks. This continued until the passage of the 1977 amendments which prohibited dispersion techniques as a method of meeting local ambient air quality standards. However, the weakness of the Clean Air Act is that it emphasizes local emission standards and does not look at the impact of total emission loadings on the environment (Parker and Blodgett 1985).

It was believed that the Clean Air Act would lead to a consistent decline in SO_2 emissions over time as older power plants, which lacked scrubbers, were replaced by newer power plants with scrubbers or other pollution abatement equipment. As previously discussed, however,

Table 37
Emissions of Sulfur Dioxide and Nitrogen Oxide in the U.S., 1980 (1000 tons per year)

State	SO2	NOx	State	SO2	NOx
Alabama	760	450	Montana	160	120
Alaska	20	55	Nebraska	75	190
Arizona	900	260	Nevada	240	80
Arkansas	100	220	New Hampshire	90	60
California	445	1,220	New Jersey	280	400
Colorado	130	275	New Mexico	270	290
Connecticut	70	135	New York	950	680
Delaware	110	50	North Carolina	600	540
District of			North Dakota	100	120
Columbia	15	20	Ohio	2,650	1,140
Florida	1,100	650	Oklahoma	120	520
Georgia	840	490	Oregon	60	200
Hawaii	60	45	Pennsylvania	2,020	1,040
Idaho	50	80	Rhode Island	15	40
Illinois	1,470	1,000	South Carolina	330	200
Indiana	2,000	770	South Dakota	40	90
Iowa	330	320	Tennessee	1,100	520
Kansas	220	440	Texas	1,270	2,540
Kentucky	1,120	530	Utah	70	190
Louisiana	300	930	Vermont	7	25
Maine	100	60	Virginia	360	400
Maryland	340	250	Washington	270	290
Massachusetts	340	250	West Virginia	1,100	450
Michigan	900	690	Wisconsin	640	420
Minnesota	260	370	Wyoming	180	260
Mississippi	280	280			
Missouri	1,300	570	U.S. TOTAL	26,500	21,220

Source: Office of Technology Assessment, Acid Rain and Transported Air Pollutants, 1985, 58.

electric utilities have been extending the operating lifetimes of their power plants and some estimates suggest an increase in SO_2 and NO_x emissions over the next twenty years (Trisko 1987). This has led to initiatives to address the acid deposition directly because the Clean Air Act will not provide immediate relief to the problem.

ACID RAIN LEGISLATION*

Attempts to address the acid rain issue politically at the national level have focused on Congressional subcommittee debate over the cause of acid rain and the measures that can be taken to reduce its adverse impact on the environment. Acid rain proposals fall within the

---

*Much of this discussion is based upon interviews with the staffs of Senator Robert C. Byrd, Senator Jay Rockefeller, and Congressman Bob Wise, June 1988.

broader discussion of ozone levels, which are not being met in many urban areas. The Reagan administration position has been that the Clean Air Act was working but that increased funding for acid rain research and some limited control strategies should be undertaken.

The House Energy Committee has been stalemated on the clean air issue since 1982. Henry Waxman (D-Calif.), who chairs the Subcommittee on Health and the Environment, has put forth aggressive clean air proposals but has been effectively thwarted by John D. Dingell (D-Mich.) who chairs the full committee. Waxman, who is from Los Angeles, is concerned about the health impacts of air pollution, particularly ozone, whereas Dingel, from Detroit, is concerned about the health of the automobile industry. On the Senate side, Senator Robert C. Byrd (D-W.Va.), the past majority leader, would not bring acid rain control bills to the floor because of concern for the welfare of the coal and utility industry in West Virginia. Senator Byrd has been against any proposal requiring an immediate reduction in emissions. The Senator is now Chairman of the Senate Appropriations Committee, which still puts him in a powerful position to influence undesirable acid rain control legislation.

Acid rain proposals that have been discussed generally comprise the following components: (1) establishing a specific SO_2 and NO_x emission reduction achievable according to a particular timetable; (2) targeting a specific number of states in which the emission reductions will be achieved (often the 31 eastern states) and state plans to achieve these goals; (3) targeting a specific group of emitters in which reduction will occur; (4) prescribing the method in which the emitters may reduce their emission; and (5) determining how the cost of emission control will be distributed (Parker and Courpas 1988).

Most proposals call for a reduction of at least 10 million tons of SO_2 emissions (below 1980 levels) achievable sometime in the next decade. Most bills, either directly or indirectly, single out coal-fired power plants as the major source of SO_2. The top 80 utility SO_2 emitters are listed in Table 38. American Electric Power's Gavin plant, located in Ohio across the Ohio river from West Virginia, is the nation's top emitter of SO_2; and six of the top fifty are located in West Virginia. These are Harrison (11), Mitchell (18), Kammer (25), Amos (40), Fort Martin (44) and Mt. Storm (50). Only one West Virginia plant, Pleasants, is equipped with a scrubber. It is estimated that mandatory scrubbing of the top 50 emitters would cost $4.2 billion in annual levelized dollars and remove 6.1 tons of SO_2 (Trisko 1983). Concern over the life extension and operation of older coal-fired units has led to the introduction of legislation requiring power plants to meet the NSPS after they are 30 years old. It has been estimated that this approach would reduce SO_2 by more than 12 million tons at an annual cost of over $10 million (Parker and Blodgett 1987).

Some initiatives other than those in Congress address the acid rain problem. Eight northeastern states (Connecticut, Maine, Massachusetts, New Hampshire, New Jersey, New York,

Table 38
Highest Sulfur-Dioxide-Emitting Utility Plants

Plant Name and State	SO_2 Emissions (1000 tons)	(lbs/MMBtu)	Plant Name and State	SO_2 Emissions (1000 tons)	(lbs/MMBtu)
Gavin, OH	376.4	5.1	Jappa, IL	100.1	3.3
Cumberland, TN	356.8	6.5	Sioux, MO	97.2	4.5
Paradise, KY	342.7	6.8	Shawnee, KY	96.6	2.2
Gibson Station, IN	305.6	4.9	Ft. Martin, WV	95.1	3.0
Clifty Creek, IN	288.2	6.4	Tanners Creek, IN	94.9	4.2
Baldwin, IL	259.3	5.2	Mill Creek, KY	94.1	4.1
Bowen, GA	248.1	2.8	Morgantown, MD	92.0	2.8
Muskingum, OH	244.1	7.7	Mount Storm, WV	91.7	2.9
Labadle, MO	217.2	4.2	Miami Fort, OH	89.5	2.6
Monroe, MI	213.9	2.9	Avon Lake, OH	89.5	4.1
Harrison, WV	215.1	4.0	Kingston, TN	88.6	1.7
Wansley, GA	209.8	3.7	Belews Creek, NC	86.6	1.5
Kincaid, IL	205.5	6.5	Ghent, KY	85.4	3.3
Conemaugh, PA	205.2	3.8	Marshall, NC	85.1	1.6
Kyger Creek, OH	202.3	6.2	Burger, OH	84.7	6.1
Conesville, OH	200.0	5.2	Backjord, OH	84.4	2.8
Madrid, MO	198.9	5.6	Gorgas 2 & 3, AL	84.3	2.4
Mitchell, WV	171.3	4.0	Brayton, MA	83.2	2.1
Hatfield, PA	171.2	4.1	Sevier, TN	81.4	3.2
Homer City, PA	169.1	3.2	Four Corners, NM	80.7	1.1
Gaston, AL	169.1	2.8	La Cygne, KS	78.0	2.5
Montrose, MO	162.0	11.7	Northport, NY	77.6	2.2
East Lake, OH	155.0	4.9	Crist, FL	76.3	3.6
Big Bend, FL	153.2	4.6	S. Oak Creek, WI	75.7	4.7
Kammer, WV	149.1	6.7	Michigan City, IN	75.4	4.7
Keystone, PA	142.3	2.7	Allen, TN	73.9	3.7
Brunner Island, PA	139.0	3.0	Crystal River, FL	72.5	3.1
Coffeen, IL	137.3	6.9	Breed, IN	71.0	6.6
Gallatin, TN	137.1	4.5	Canal, MA	70.6	2.3
Sawmais, OH	137.0	2.5	Roxboro, NC	70.5	1.2
Johnsonville, TN	135.3	3.7	Ashtabula, OH	68.5	5.6
Cardinal, Oh	126.0	3.0	Coleman, KY	68.3	4.0
Colbert, AL	125.9	3.7	Contralia, WA	68.3	1.7
Hill, MO	121.6	8.5	Monticello, TX	68.2	1.1
Cayuga, IN	115.9	4.1	Widows Creek, AL	67.4	1.7
Stuart, OH	112.8	1.8	Wabash River, IN	66.0	3.9
Montour, PA	109.5	2.3	Watson, MS	65.9	3.6
Petersburg, IN	108.5	2.9	Campbell, MI	65.5	3.1
Yates, GA	106.6	1.1	Ashbury, MO	65.0	9.3
Amos, WV	105.1	1.2	Shawville, PA	64.7	3.2

Pennsylvania, and Vermont) are now involved in a U.S. Circuit Court of Appeals suit in Washington, D.C., against the U.S. Environmental Protection Agency because the agency has not revised the air-pollution control plans of Illinois, Indiana, Kentucky, Michigan, Ohio, Tennessee, and West Virginia to reduce emissions of SO_2. The Environmental Protection Agency did modify its regulations to prevent the use of tall stacks as a means of merely dispersing pollutants, which does have a great impact on areas such as West Virginia, eastern Ohio, western Pennsylvania, eastern Kentucky, and Tennessee in siting new facilities. It is difficult even to site "clean coal facilities," such as fluidized bed combustion plants, in narrow river valleys. Also, it is conceivable that some plants with retrofitted scrubber equipment may not be able to comply with stack height regulations (Trisko 1985).

The Center for Clean Air Policy's proposal to equip the Gavin plant with a scrubber and to sell power to New England via a new transmission line is estimated to reduce SO_2 emissions from this one plant by at least 320,000 metric tons per year (Parsons 1987, 48). Even if this approach were used on additional capacity, as suggested by the Center for Clean Air Policy, this could not lead to the level of reductions called for in most recent proposals. However, any emission reduction of this magnitude should be viewed as a positive step.

There have also been some state initiatives. Most recently, New York Governor Mario M. Cuomo and Ohio Governor Richard F. Celeste announced a proposal on June 6, 1988, to reduce emissions of SO_2 by 10 million tons a year (from utility sources) by the year 2003 with an authorization of $900 million annually to finance up to half of the capital costs of these controls. In addition, the proposal would expand the clean coal technology program. Part of the cost of this proposal would be financed by a fee on oil imports, which would be used to increase the Strategic Petroleum Reserve (State of New York, Press Office 1988). This proposal was attacked by other members of the Ohio delegation, and little support for this proposal is expected from the coal state delegations.

Lacking movement in Washington on this issue, many states have passed their own acid rain legislation. This includes New York's program to reduce in-state emissions, Wisconsin and Minnesota's emission caps, and California's multicounty region in the Los Angeles area to attack the acid fog problem (Sullivan 1984, 23-29). Such independent action strengthens the position of those interests when more stringent controls are requested at the national level.

CANADIAN-AMERICAN RELATIONS:
THE TRANSBOUNDARY POLITICS OF ACID RAIN

Because acid rain has been chiefly identified as a long-range transport pollution problem, it has become a major bilateral issue between the United States and Canadian governments. Considerable confusion exists in both countries as to what is or is not being done. In addition to diplomatic haggling on this issue, economic interests in the United States claim that the Canadians are using the acid rain issue as a means to increase their electricity sales to the United States (Friedman and McMahon 1984). The relevancy of these issues and their resolution could have important implications to the development of a coal-by-wire strategy for the Appalachian region.

The transboundary politics of acid rain between the United States and Canada can best be divided into two periods. The first period began in 1978, when the United States and Canada entered into diplomatic negotiations that were to lead to the signing of an international air quality agreement, and ended around 1985, when the Canadian government in Ottawa passed an aggressive two-part acid rain control program. The second period started with the passage of the Canadian 1985 act and continues to the present.

The first phase began with the creation of numerous U.S.-Canadian Task Forces organized to deal with tolerance levels within the environment, determining remedial actions, defining technical and socioeconomic costs of control, and examining legal and institutional questions. In 1983 the final report of the Canada-U.S. Work Groups was made. The crux of this report probably best exemplifies the Canadian government's official position. The work groups agreed on the following points:

a. Damage in both the short and long terms is occurring in areas vulnerable to acid rain as a result of sulfur deposition;

b. Wet sulphate deposition above 18 lb/acre/yr in vulnerable areas is associated with damage; areas with deposition less than this level have no recorded damage;

c. The damage is caused by sulfur deposition, and the solution is to reduce it;

d. Acid rain occurs in eastern North America in and downwind from major industrial regions;

e. Technology exists to reduce emissions by substantial amounts;

f. If there are no changes in abatement programs, the emissions are forecast to increase through the remainder of the century.

The joint report indicated that U.S. annual SO_2 emissions amounted to 30 million tons as opposed to 5.3 million tons emitted from Canadian sources. The joint report admitted that both countries contribute to one another's acid rain problem but emphasized that American emission exports come primarily from the American Midwest (Environment Canada, 1984, 6).

While the diplomatic discussions were going on, Canada set new precedents in Canadian diplomatic behavior. These acts included direct outreach to the American people, increased Congressional lobbying, provincial involvement in U.S. administrative rule making, and financial support of environmental lobbying in Washington D.C. These efforts no doubt reflect the importance of the acid rain issue to the Canadians (Carroll 1982).

While the government in Ottawa has sought to negotiate with the United States, its own bargaining position is weakened by a federal system that gives the provincial governments responsibility for most environmental regulation. As such, Ontario is the centerpiece of much of the Canadian-American controversy.

Ontario is highly vulnerable to acid deposition yet is a significant source of SO_2 and NO_x. Up through 1982, the provincial government of Ontario took action to impose voluntary steps to reduce emissions from Ontario Hydro, to increase funding for acid rain research, and to financially support groups to lobby in Washington.

What diplomatic good Ontario's action may have done was hurt by a January 1982 decision by Ontario Hydro to sell coal-fired electricity to states in the Northeast. This action, coupled with increased electricity sales in general to the Northeast throughout the 1980s, has sparked widespread speculation that acid rain is simply a ruse intended to hurt the electric power industry in the Midwest, which competes in the same electric power market.

As mentioned earlier, the inability of the federal government in Ottawa to get provincial cooperation in addressing acid rain issues in Canada has hindered Canadian efforts to move the United States toward the Canadian solution. This condition existed through 1985. At that time, the Canadian Federal Government pushed forward a two-part acid rain control program with the overall objective to reduce deposition of wet sulfate to no more than 20 kilograms/hectares/year. It was hoped that such action would spur parallel cuts by the United States. (Courpas and Parker 1988).

These standards are to be met by 1994 and are based upon reducing SO_2 emissions from a 1980 base year. In that year the seven eastern Provinces emitted 3,828,000 tons of SO_2. In 1986 total emissions of SO_2 were 3,080,000 and the 1994 legal limit has been set at 2,424,000 tons. It has been estimated that over 75 percent of these new reductions will come from installing new pollution control equipment. All of the eastern provinces have entered into agreements with the federal government in meeting these requirements. This second stage of Canadian acid rain policy is built more upon the idea that Canada should clean up its own house. Under this plan, the bulk of these reductions will come from Ontario, Canada's largest provincial emitter.

In addition to setting targets for reduced SO_2 levels, Canadian automobiles and light-duty trucks beginning with model year 1988 have to meet U.S. standards for NO_x, CO, and HC (Courpas and Parker, 1988).

Evidence seems to point clearly that the Canadian government has made real efforts to reduce their own emissions of SO_2. Still, a comparison between United States and Canadian efforts is difficult to make. Table 39 compares the two countries in terms of their relative emissions of SO_2 and NO_x from 1980 to 1986. As can be seen, both countries have made progress in reducing SO_2 emissions during this time period. Canadian NO_x emissions have risen slightly, but new pollution standards for automobiles should help reduce these in the near future.

Table 39
U.S. and Canadian SO_2 and NO_x Emissions, 1980-1986 (millions of metric tons)

Year	SO2		NOx	
	U.S.	Canada	U.S.	Canada
1980	23.2	4.6	20.3	1.8
1981	22.4	4.4	20.3	---
1982	21.3	4.6	19.5	---
1983	20.5	n/a	19.1	---
1984	21.3	3.9	19.7	---
1985	21.6	n/a	19.7	---

Sources: U.S. Environmental Protection Agency, 1988.
Canadian Ministry of the Environment, 1988.

A major distinction between the two countries' pollution control strategies is that the Canadians set absolute emission loadings for each province. Since there is an emission ceiling, any new source that would go beyond this limit would have to be offset at some other point source of pollution. In contrast, the Clean Air Act sets rates for individual sources and is not so much concerned with total pollution loadings. Under these conditions, it is possible that one can be regulating pollution under the law yet have total emissions increase due to increases in the number of polluters. Actions by both provincial and the federal government in Canada suggest that acid rain is more than a ruse to sell electricity.

The greatest potential for increasing Canadian imports of electricity probably will come through hydro and nuclear power developments. In a 1982 study by the Congressional Research Service, nuclear power was identified as the most likely source of substituting future coal-fired capacity. The report noted that Canadian nuclear power developments like their American counterparts are subject to long lead times and a difficult regulatory review (Courpas and Parker 1988).

In respect to hydro power developments, the present Canadian situation is rather clouded. Much of Canada's existing hydro capacity is expected to be utilized in Canada. What surpluses exist have been identified as available only for short-term exports. However, in Quebec, considerable attention has been given by the provincial government to increasing Quebec's role as an energy exporter.

In a report conducted by the National Coal Association in 1987, it was found that through 1985, imports of Canadian electricity into the eastern part of the country

were directed at displacing higher cost oil and gas generation; in the midwest, at displacing higher cost, older coal-fired generation; and in the western-most states, at meeting electricity generation shortfalls caused by poor hydro electricity production.
Since this report was released, the United States and Canada have negotiated a free trade agreement. The agreement addresses several energy issues that exist between the two countries, and it has important implications for increased electricity sales between the two countries. The agreement basically sets Canadian electricity to a market price as opposed to the present pricing mechanism which had been based upon the least-cost alternative to the U.S. utility purchasing the power. In addition to changes in the pricing mechanism, questions of supply and energy security have been answered by Canadian assurances that U.S. energy consumers will be treated on an equitable basis with Canadian consumers, and if supply problems should arise available energy will be shared proportionately between domestic and export markets.
Canadian sales of electricity represent approximately 2 percent of the total electricity produced in the United States. Their importance varies from region to region:
New England imports 9 percent of its power; New York, 12 percent. No doubt these figures will increase in the future, but the free trade agreement leaves uncertain the magnitude of that increase (National Coal Association 1988).
In a speech in 1987, Minister of Energy, Mines and Resources Marcus Massey stated that, "Canadians are right to try to expand their electricity exports." Mr. Massey gave the following as reasons: (1) creating jobs; (2) creating technologies and increasing engineering experience; (3) raising national income; and (4) paying for Canada's own energy needs (National Coal Association 1988).
Not all Canadians want to build power plants or hydroelectric dams and sell the power to the United States. In fact, the free trade agreement and its energy components have been quite controversial in Canada. Environmentalists and native Indians, to name a few, are concerned about projects such as the James Bay II project and hydro developments in the west that have export potential. Many Canadians resent the idea of Canada's becoming an American energy colony.
In summation, Canada has made real efforts in reducing SO_2 and NO_x. Ceiling levels for such emissions make the Canadian law increasingly stringent over time. Canadian imports of power will no doubt increase to some degree in the future, but this is probably the result of market conditions, the current status of America's nuclear industry, and the utility industry in selected areas of the country.

CONCLUSION

Although there are many who maintain that the environmental impacts of acid rain are greatly overstated, it is obvious that the issue has generated sufficient, deeply rooted concern which will result in action. The timing of acid rain control measures and the deadlines and emission targets do remain in question, but they will affect the economic impact of legislation to the Midwest and Appalachians. Any substantial reduction targets with short-term deadlines (less than five years) will be most difficult to attain without retiring some capacity. Also, for those power plants that are equipped with scrubbers, the cost of power production will rise reducing the ability of the area to sell economy energy to the East. On the other hand, new power plants could be built in the Midwest and Appalachians which would meet strict standards and be competitive with load center facilities. Competition from Canada exists, and Canada will be placing great pressure on the United States to institute acid rain controls in light of its own recent reduction effort.

BIBLIOGRAPHY

Canadian Embassy. 1984. Fact Sheet on Acid Rain. Washington, D.C.: Canadian Embassy.

Cannon, James S. 1984. Acid Rain and Energy: A Challenge for New Jersey. New York: INFORM.

Carroll, John. July 1982. "Acid Rain: An Issue in Canadian-American Relations, Canadian-American Committee." Washington D.C.: Howe Institute and National Planning Association.

Congressional Digest. 1985. "The Acid Rain Controversy." Washington, D.C.

Courpas, Mira, and Larry Parker. April 20, 1988. Canada's Progress on Acid Rain Control: Shifting Gears or Stalled in Neutral? Congressional Research Service Report.

Friedman, James M., and Michael S. McMahon. 1984. The Silent Alliance. Chicago: Regnery Gateway.

General Accounting Office. 1984. An Analysis of Issues Concerning "Acid Rain." RCED-85-13. Washington, D.C.: U.S. Government Printing Office.

Hunter, S., T. Basile, G. Sayre, and F. Calzonetti. 1988. An Analysis of Acid Rain Policy for West Virginia. Morgantown, W.Va: West Virginia University, Energy and Water Research Center Report, 1988.

Keenan, Boyd. August 30-September 2, 1984. "Acid Rain Policy: Comparing Canadian and U.S. Approaches." Paper presented at the 1984 annual meeting of the American Political Science Association, Washington, D.C.

National Coal Association. April 13, 1988. Statement of Thomas Altmeyer, Sr. Vice President, Government Affairs National Coal Association on the Canada-U.S. Free Trade Agreement before the Finance Committee United States Senate.

---. July 1987. The Effects of Imports of Canadian Electricity on the U.S. Coal Industry 1980-1985. Washington, D.C.: National Coal Association.

Navarro, Peter. 1985. The Dimming of America, Cambridge, MA: Ballinger.

Office of Technology Assessment. 1985. Acid Rain and Transported Air Pollutants: Implications for Public

Policy. Washington, D.C.: U.S. Office of Technology Assessment.
---. 198?. <u>The Effects of Acid Rain Control Measures on the Coal Market</u>. Washington, D.C.: U.S. Office of Technology Assessment.
Parker, Larry. September 15, 1982. <u>Acid Rain Legislation and Canadian Electricity Exports: An Unholy Alliance?</u> Washington, D.C.: Congressional Research Service.
Parker, Larry, and John Blodgett. August 4, 1987. <u>Acid Rain: Issues in the 100th Congress.</u> Washington, D.C.: Congressional Research Service.
---. 1985. "Acid Rain Legislation and the Clean Air Act: Time to Raise the Bridge or Lower the River?" In <u>The Acid Rain Debate</u>, edited by Ernest J. Yarnella and Randal H. Ihara. Boulder, Colo.: Westview Press.
Parker, Larry, and Mira Courpas. 1988. <u>House Acid Rain Bills on the 100th Congress.</u> Washington, D.C.: Congressional Research Service.
Parsons, Edward A. 1987. <u>Midwest-Northeast Transmission: A Partial Solution to Acid Rain?</u> Unpublished report, submitted to the Center for Clean Air Policy, Washington, D.C.
State of New York, Press Office. June 6, 1988. "Acid-Rain/Energy Security Initiative."
Strahler, Arthur N., and Alan H. Strahler. 1983. <u>Modern Physical Geography</u>. 2d ed. New York: John Wiley and Sons.
Sullivan, Carol. 1984. "Acid Rain . . . Blowin' in the Wind." <u>State Legislatures</u>: 23-29.
Trisko, Eugene M. 1987. "Resolving the Clean Air Act Impasse." <u>Future Resources</u> (December): 22-25.
---. 1984. "The Clean Air Act and Acid Rain." <u>Electric Perspectives</u>. Washington, D.C.: Edison Electric Institute.
---. November 15-17, 1983. "Potential Impacts of Acid Rain Control Legislation." Paper presented at the 6th International Coal and Lignite Utilization Exhibition and Conference, Houston, Texas.

8

Siting Energy Facilities

Frank J. Calzonetti

Siting power plants and transmission lines has generally become a more difficult problem for energy planners. In the past, the siting process was less complicated, and projects could be completed in less than five years. A power company would persuade the state public utility commission that the new facility was needed, receive a certificate of public convenience and necessity, then proceed with the construction of the new power plant or transmission line. Starting in the late 1960s, new environmental and health legislation was passed both at the federal and state levels. Such legislation usually required additional permits and often called for public hearings to review proposals, processes which stretched out the time before the project could be completed. In addition to the increased demands of new environmental legislation, public interest groups became much more active in expressing concern about development. Power plants and transmission projects often were the focus of opposition. Court hearings and other protests delayed many projects and halted others. While many power plants and transmission projects were plagued by delay, others were sited on time without much notice.

We believe that the level of opposition to energy projects depends partly on the perception of the energy technology, the familiarity of local residents with the technology, and the benefits to the community that are perceived to result from the siting of the facility. Nuclear power is now almost uniformly unacceptable to most Americans, and new nuclear facilities will encounter problems anywhere they are located. National organizations have antinuclear positions and will fight nuclear proposals irrespective of their location. The acceptability of coal depends on where the facility is to be located and the technology used. A coal plant proposed in an area not accustomed to such facilities is less likely to be supported than one in a coal-producing area. Not only will the coalfield residents be more familiar with the

the technology, but the economic impacts, due to coal mining linkages, will be greater in the coal-producing region. A new coal technology, such as a coal liquefaction facility, may not be supported since the residents may view the facility as noxious and not worth its economic benefits.

In this chapter we review the nature of permits needed to site new power plants and transmission lines. We show that the permits will vary across states and according to the energy developer. We then turn to a discussion of the public acceptance of new power plants. We provide two recent case studies of siting coal-fired power plants. In one case, the project was withdrawn because of public opposition; in the other, the project went very smoothly.

PERMITS FOR POWER PLANTS

General Requirements

The number and type of permits needed to site a new power plant depend upon the type of facility and its location. All power plants require permits to satisfy federal and state laws and regulations. There is considerable variation in the siting process across the states, but most states require submission of documentation showing the plant is needed and an environmental report. These are subject to public hearings and rulings before permits are issued to allow the project to go forward. Nuclear power plants call for a much greater federal involvement, through the Nuclear Regulatory Commission, than fossil or hydroelectric plants. In most situations, local permits may be required as well. In this section we review the nature of the permits required to site a new coal-fired power plant.

In most states, utility-sponsored projects will deal mainly with state agencies, but federal approvals are almost always required. As mentioned earlier, the U.S. Army Corps of Engineers is often the lead agency for projects in the eastern United States, and the U.S. Bureau of Reclamation assumes this role in the western states. Because a federal permit is required, an environmental impact statement (EIS) must be filed as required by the National Environmental Policy Act (NEPA). In most states, authority to issue air and water permits have been approved by the federal government. The U.S. Environmental Protection Agency is not usually the lead federal agency, but it does review the EIS. In addition, the Federal Aviation Administration (FAA) and Coast Guard also issue permits for construction and operation activities of the new power plant.

Utilities must obtain a Certificate of Public Need and Necessity from the state public utility commission to add the new capacity to the rate base. Public hearings are held before such a certificate is issued. Many environmental hearings may also be held. In addition to demonstrating the need for the facility, and that there are no alternatives that offer less expensive or more reliable service, the utility must show that the proposed power plant will meet standards regarding the protection of the environment. Areas of environmental concern include air, water, land, historical sites, wildlife, and aesthetics. Air emissions are the principal concern for coal-fired power plants sited in the Appalachians. Of major concern are sulfur oxides and particulates, however, nitrous oxides, carbon monoxide, and carbon dioxide are also emitted by coal-fired power plants. Air quality regulation falls under the purview of the Air Pollution Control Board in West Virginia which implements the federal Clean Air Act standards. Table 40 lists the air quality standards that must be met. The Air Pollution Control Board does not have a permit form which must be submitted. Information concerning the power plant must be provided in writing to the agency in the form of a detailed technical document that allows adequate review. A model of stack emissions must also be provided to determine whether ambient air quality standards will be violated in proximity to the site. Furthermore, impacts on the Prevention of Significant Deterioration (PSD) must also be evaluated.

The time frame to obtain approval of a permit is quite long. From a statutory perspective, there is a ninety day clock which is initiated at the receipt of the completed permit application. This clock is reset when and if a permit application is deemed deficient in information within the first thirty days. Frequently up to one year of ambient air quality data must be collected and analyzed before the application is filed with the Air Pollution Control Commission. As a result, a typical permit may take as much as two years from initiation to approval or denial. An application for a state-owned authority would not be able to change this time frame.

The maintenance of water quality in West Virginia is delegated to the Water Resources Division of the Department of Natural Resources (DNR). The application must demonstrate that its discharge will not violate water quality standards.

Solid waste is another area of concern, particularly since new plants require flue gas desulfurization equipment. Substantial amounts of fly ash and sludge are generated and must be dispersed nearby. The construction of power plants may also present unacceptable problems because of the destruction of wildlife habitat. There are many ways in which habitats can be destroyed by constructing power plants, or by any other large project, for that matter. The physical loss of the site could result in the loss of habitat. Power plants are often located in floodplain areas which provide good habitat for wildlife. Impediments or temperature changes to water may alter the existing ecosystem. Such habitat considerations must be carefully considered when there are threatened or endangered species in the area of the site. Loss of habitat could be a significant problem. Numerous studies indicate that sulfur oxide emissions may affect a number of plant and crop species frequently found in West Virginia. Another concern is aesthetics. Power plants dominate the local landscape and may have aesthetic impacts unacceptable to local residences. In West Virginia this is less of a

Table 40
West Virginia Air Quality Standards (in ug/cubic meter)

	Annual Mean	Primary 24-hour Maximum	Secondary 3-hour Maximum
Particulates	75	260	150
Sulfur Oxides	80	365	1300
Nitrous Oxides	100	---	---
Hydrocarbon	---	---	160 (primary)
Carbon Monoxide	Maximum 8-hour concentration		10
	Maximum 1-hour concentration		40
Ozone	Maximum 1-hour concentration		235

Note: All standards except ozone are not to be exceeded more than once per year; ozone is not to be exceeded more than once per year on a three-year average.

Source: West Virginia Administration Regulations, chaps. 16-20, series VIII, IX, and XII,

problem since power plants are often concealed in river valleys, even though the steam plume may be visible for many miles (depending upon weather and operating conditions). Noise is also an impact for those living nearby. In our survey of communities adjacent to power plants, the noise of power plants was often noted by those close to the facility.

Mountaineer Plant, New Haven, West Virginia

In 1974, Appalachian Power Company, a subsidiary of the American Electric Power System, announced plans to construct a 1,300-megawatt coal-fired power plant in Mason County, West Virginia. The plant's major components consist of one building housing one boiler and one 1,300-megawatt generating unit; one cooling tower; one river make-up structure; one 1,100-foot high stack, one set of electrostatic precipitators; one bottom ash/waste water treatment facility; coal-handling facilities (to accommodate both rail and water delivery); one 765-kilovolt switchyard; four transmission lines; tanks to hold condensate, fuel oil, and water; and a material unloading lock (U.S. Army Corps of Engineers 1977, 1). The plant was to require 440 acres, the switchyard 37 acres, and the fly ash disposal area approximately 500 acres (U.S. Army Corps of Engineers 1977, 15). Completion of the plant was scheduled to begin in the summer of 1979. The project was completed without much public comment and went into operation in 1980.

Table 41 lists the permits and approvals necessary to construct this plant. Since the plant required U.S. Army Corps of Engineers permits to build structures on a navigable river, the corps was designated as the lead federal agency and filed an EIS. The draft EIS was completed in 1975, and the final EIS was submitted in 1977. Table 41 indicates that most of the permits for this power plant were secured from West Virginia state agencies.

Current Trends

In many states the developer must contact many state agencies to obtain the necessary permits and approvals. In the 1970s, some states instituted streamlined siting procedures to simplify the licensing process and perhaps expedite the siting process. Many utility representatives claim that overlapping and redundant permits and approvals extend the licensing process causing project costs to rise. The Kaiparowit power project in Utah, for example, was initially proposed in 1961 as a 5,000-megawatt complex which would cost $500 million. By 1976, the project was still not completed, but its price tag had risen to $3.7 billion even though the utility sponsors reduced the size of the project to 3,000 megawatts (Myhra 1977, 25). According to the utilities involved in the project, the project was cancelled because of red tape and public opposition. No mention was made of the lack of the need for the facility in the face of downward revisions of electricity demand growth (Rock 1977, 250). In any case, many states did implement siting laws to streamline the siting process. It is not clear whether such programs achieved this goal.

In the case of the Mountaineer plant, Appalachian Power needed to contact many different agencies but had little difficulty in completing the project on schedule. Notwithstanding, in 1985, Governor Arch Moore, of West Virginia did consolidate permitting in the state Department of Energy. Although this may ease the permitting process for the utilities, it is doubtful whether this will quicken siting since it will not eliminate the need for on-site environmental monitoring and studies of proposed sites.

As discussed earlier, however, the governor's Public Energy Authority has the power to construct state-owned power plants and has the right of eminent domain. It can also issue tax exempt or taxable bonds, commercial paper, or other evidences of indebtedness. Since the Public Energy Authority is not a utility, it can build power plants that need not obtain a certificate of need and necessity from the West Virginia Public Service Commission. This may hasten siting in a time when the state Public Service Commission may be reluctant to issue permits for new capacity expansion for electric utilities because of excessive capacity reserves and low forecasts of demand increases.

Table 41 shows that local permits were not required to construct this new power plant. This is often the case in West Virginia which lacks county zoning in most areas. In

Table 41
Permits Required for Siting a Coal-Fired Power Plant in West Virginia

I. Federal

    A. U.S. Army Corps of Engineers
       1. Make-up water intake structure
       2. Coal Unloading Dock
       3. Unloading Dock
       4. Diversion of Little Broad Run

    B. Federal Aviation Administration
       1. Stack - obstruction lights
       2. Cooling Tower

    C. U.S. Coast Guard
       1. Obstruction warning lights on river structures
       2. Registration of tug boat
       3. Oil transfer operations manual

    D. U.S. Environmental Protection Agency
       1. National Pollution Discharge Elimination System

II. State of West Virginia

    A. Air Pollution Control Commission
       1. Approval to construct air emission source

    B. Department of Health
       1. Permit to construct sewage treatment system
       2. Permit to discharge water pollutants (i.e., treated sanitary sewage)
       3. Permit to install potable and sanitary treated water system
       4. Interior storm water system
       5. Interior sanitary system
       6. Landfill type fly ash disposal area

    C. Department of Natural Resources
       1. Permit to construct, operate and discharge from the combined bottom ash/wastewater pond and adjacent recirculation pond
       2. Disposal of industrial wastes in combined bottom ash/wastewater pond
       3. Disposal of industrial wastes in the landfill type fly ash disposal area
       4. Approval of plans for wastewater treating facility
       5. Proposed access to plant site from state road

    D. Department of Highways
       1. Acceleration and deceleration lanes at state road
       2. Pipesleeves under state road
       3. Power lines over state road
       4. Traffic signal lights
       5. proposed access to plant site from state road

    E. Department of Labor
       1. Permit to install and operate elevators
       2. Certificates of steam boiler inspection (main and auxiliary boilers)

    F. Fire Marshall Office
       1. Approval of plans to install fuel oil tank

    G. Environmental Protection Agency
       1. Permit to drill water wells

    H. Department of Industrial Relations
       1. Building permits
       2. Permit to install electrical facilities in each building or structure
       3. Permit to erect pumphouse for circulating ash water basin
       4. Permit to erect fire pump building and water tank
       5. Permit to construct temporary office and storage building

    I. Public Service Commission
       1. Certificate for Public Need and Necessity

III. Local Permits: None Required

Source: U.S. Corps of Engineers, 1977. *Final Environmental Impact Statement, Project 1301*, 1977, pp. 324-329.

other parts of the country, local ordinances and taxes can have a significant role in the siting process.

TRANSMISSION LINE SITING

Like power plants, utilities in most states must submit evidence to state public utility commissions that new transmission facilities are needed and documentation that the line will not cause undue harm to the environment. Public hearings on the need and impact of transmission lines may be held, and it is necessary to secure the required permits before construction can begin. The National Governor's Association Electricity Transmission Task Force completed a survey of licensing requirements for new transmission lines and found that state public utility commissions were most often responsible for approving projects (National Governor's Association 1987). Other offices involved in transmission projects include state energy offices, energy facility siting boards, state environmental offices, and state land use offices (National Governor's Association 1987, 9). The task force found that a great deal of variety exists in the permitting process and that the time between initial filing and final approval can be as short as two months or as long as three years (National Governor's Association 1987, 9). Delays from public opposition or local community objections may prolong the process well beyond three years.

Table 42 lists the federal, state, regional, and local offices to be contacted in the siting of a 50-mile transmission line in a state in the East Central Area Reliability Council (ECAR) region. Depending upon the region, federal involvement can be quite extensive. Lines crossing federally owned land, areas designated as the critical habitat for endangered species, wilderness areas, coastal areas, wetlands, or Indian reservations require federal permits.

The potential health effects of high-voltage transmission lines is a concern that must be considered in any plans to site new power lines. The oscillating electric fields resulting from the operation of power systems at a frequency of sixty hertz can cause biological effects depending upon the strength of the field. Numerous experimental and epidemiological studies have investigated the biological effects of electric fields. In widely publicized epidemiological studies, Wertheimer and Leeper report associations between chronic exposure to electric distribution lines and childhood leukemia and other cancers (Morgan et al. 1985; Morgan et al. 1988). These studies compared houses, rather than people, and did not adequately control for other intervening variables. Most studies, however, that do show serious biological effects on humans or animals cannot be replicated consistently. It seems that electric fields can interact with biological systems through mechanisms involving the surface of cells, affect the rate at which calcium ions bind to neural tissue, and affect the rate of certain hormone production (Morgan et al. 1988). Whether these effects result in disease, death or disability awaits further study.

Table 42

Federal, State, Regional, and Local Contacts Required for a Specific 50-mile Transmission Line Application within One State in ECAR

(This is an illustrative example, other projects could involve a greater or fewer number of contacts.)

Federal Agencies
U.S. Geological Survey
U.S. Government Printing Office
U.S. Fish and Wildlife Service (2 locations)
Federal Aviation Administration (6 locations)
U.S. Department of Agriculture (3 locations)

State Agencies
State Siting Agency
State Department of Natural Resources
   Division of Forestry
   Division of Lands and Soils
   Division of Natural Areas and Preserves
   Division of Oil and Gas
   Division of Oils and Water Conservation
   Division of Geological Survey
   Office of Public Information and Education
   Office of Real Estate
   Office of Parks & Recreation
State Department of Transportation
   Division of Aviation
   Bridge Inventory Department
   District Office (2)
   Bureau of Utilities & Properties
State Environmental Protection Agency
   Water Quality Section
   Air Quality Section
State Department of Agriculture
   Historic Preservation Office
   Crop Reporting Service
   Agricultural Research and Development
State Department of Economic & Community Development

Regional & Local Agencies
Areawide & Regional Coordination Agencies (2)
County Offices of 5 Different Counties
   Planning Commission
   Park Department
   Health
   Commissions Office
   Administrators
   Engineer
County Township Offices of 24 Different Townships
City & Village Offices (4)

Source: ECAR/MAAC Coordinating Group, ECAR/MAAC Interregional Power Transfer Analysis, 1985, p. 74.

In any case, action has been taken against power line projects because of the anticipated health effects. In many hearings throughout the country, farmers have complained of impacts to cattle and poultry because of power lines. The field underneath the line can be strong enough to be felt by humans and animals so it is not surprising to hear that cattle cannot be moved across pastures beneath lines. Six states (Montana, Minnesota, New Jersey, New York, North Dakota, and Oregon) now have field limits at the boundary of right-of-ways that can not be exceeded. In New York State, a class action suit was filed by land owners because they expected the health effects of the line to reduce their property values. New York State officials are sensitive to the health concern issue with regard to power line projects and there may need to be some resolution of the issue before new power line projects can proceed.

Whereas power plants impact one site, but provide important tax benefits and employment opportunities to a local area, transmission lines are linear facilities that often provide little obvious benefit to communities or landowners along the line. In most states, localities must be consulted when power lines are proposed. In twelve states, localities can disapprove the construction of lines through their jurisdiction even though the line may provide greater statewide benefits. Local opposition of this type has stalled the 500-kilovolt loop around Washington, D.C., and Baltimore, which was approved at the state level. If the states involved could have preempted local authority, as is the case in some states, this loop might, by now, have been completed, thereby greatly increasing power trade from the Appalachians and Midwest to the East.

Communities not served by the line may not be willing to approve a facility which provides electricity for distant customers. This is also true for landowners. Public utility commissions are concerned about the distribution of costs and benefits associated with transmission lines. The balance of costs and benefits can also be a problem when an intermediary state is crossed in a transmission project. For example, transmission concepts to construct lines from West Virginia to New Jersey or New England would cross the state of Pennsylvania. In order for the state to grant eminent domain necessary for transmission line construction, the Pennsylvania Public Utility Commission must approve the project. The commission would need to be convinced that Pennsylvania residents are benefitting from the transmission project before any such authority is conferred.

An example is a line that was proposed to be extended from central Pennsylvania to serve customers in New York State. As originally proposed, the line was not to serve customers in Pennsylvania. The state would not allow the line to be built, however, unless it could be demonstrated that the project would benefit Pennsylvania residents. In order to fulfill this requirement, a substation was built in northern Pennsylvania which provided power to some local communities.

In recognition of the difficulties in siting transmission lines, the National Governor's Association made recommendations to simplify the transmission line siting process, to facilitate transmission planning and development, to coordinate multistate transmission planning, to modify rate regulation to promote transmission development, and to remedy disputed interstate transmission projects.

Pennsylvania is cast as the major impediment preventing the construction of new transmission lines from the Appalachians and Midwest to the East. It is true that there is limited transmission capacity across Pennsylvania which limits west to east transfers at some times, and that Pennsylvania state officials are not convinced that new transmission lines will provide benefits to their citizens. It is also the case that Pennsylvania officials are open to discussion and are willing to review proposals that are submitted to them. No concrete proposals have been submitted which would provide a basis for a detailed evaluation of the worth of a transmission line across the state.

The Pennsylvania Public Utility Commission regulations regarding electricity transmission lines are described in 52 Pennsylvania Code Subchapter G, §57.71 through §57.77. Table 43 lists the information that must be contained in an application from a public utility to construct a high-voltage transmission line. This information allows the public utility commission to assess the nature of the facility, the need for the facility, the location of the facility, the impacts of the facility, and alternatives to the proposed transmission line. This application is not necessary for lines to be located within existing right-of-ways or lines for which the voltage is proposed to be increased above present levels, as long as it does not change the existing right-of-way.

After receiving the application, the public utility commission arranges hearings and notifies interested individuals or agencies. The purpose of the hearing is to consider the following issues: (1) the need of the line to serve the public; (2) the safety of the line; and (3) efforts to minimize impacts to land use, soil, plant and wildlife habitats, terrain, hydrology, landscape, archaeological areas, geologic areas, historical areas, scenic areas, wilderness areas, and scenic rivers. The public utility commission will not grant an application for the proposed line unless it is determined that (1) there is a need for the new line; (2) it will not create an unreasonable risk of danger or health; (3) it is in compliance with statutes and regulations protecting natural resources; and (4) it will have minimum adverse environmental impact, considering the power needs of the public, available technology, and available alternatives (Commonwealth of Pennsylvania, Pennsylvania Code 52 §57.72 through §57.76, 1983).

The major question concerning new transmission lines across Pennsylvania is not the licensing requirements, but how the commission would view the need for a line that would provide greater service to consumers in other states than in Pennsylvania. Currently, the state of Pennsylvania has generating capacity in excess of short-term needs. Projected peak loads for 1984 are estimated to be 21,984

Table 43
Information to Be Supplied with an Application to Locate a High-Voltage Transmission Line in Pennsylvania

(1) the name of the applicant and the address of its principal business office;
(2) the name, title, and business address of the attorney of the applicant and the person authorized to received notice and communications with respect to the application if other than the attorney of the applicant;
(3) a general description-not a legal or metes and bound description-of the proposed route of the HV line, to include the number or route miles, the right-of-way width, and the location of the proposed HV line within each city, borough, town and township traversed;
(4) the names and addresses of all known persons, corporations, and other entities of record owning property within the proposed right-of-way, together with an indication of all HV line right-of-ways acquired by the applicant;
(5) a general statement of the need for the proposed HV line in meeting identified present and future demands for service, of how the proposed HV line will meet the need, and of the engineering justification for the proposed HV line;
(6) a statement of the safety considerations which will be incorporated into the design, construction, and maintenance of the proposed HV line;
(7) a description of any studies which had been made as to the projected environmental impact of the HV line as proposed and of the efforts which has been and which will be made to minimize the impact of the HV line upon the environment and upon scenic and historic areas, including but not limited to impacts, where applicable, upon land use, soil and sedimentation, plant and wildlife habitats, terrain, hydrology, and landscape;
(8) a description of the efforts of the applicant to locate and identify any archaeologic, geologic, historic, scenic, or wilderness areas of significance within two miles of the proposed right-of-way and the location and identity of such areas discovered by the applicant;
(9) the location and identity of any airports with two miles of the nearest limit of the right-of-way of the proposed HV line;
(10) a general description of all reasonable alternative routes to the proposed HV line, including a description of the corridor planning methodology, a comparison of the merits and detriments of each route, and a statement of the reasons for selecting the proposed HV line route;
(11) a list of the local, State, and Federal governmental agencies which have requirements which must be met in connection with the construction or maintenance of the proposed HV line and a list of documents which have been or are required to be filed with those agencies in connection with the siting and construction of the proposed HV line.
(12) the estimated cost of construction of the proposed HV line, and the projected date for completion;
(13) the following exhibits:
  (i) a depiction of the proposed route on aerial photographs and topographic maps of suitable detail;
  (ii) a description of the proposed HV line, including the length of the line, the design voltage, the size, number, and materials of the conductors, the design of the supporting structures and their height, configuration, and materials of construction, the average distance between supporting structures, the line to structure clearances and the minimum conductor to ground clearance at mid-span under normal load and average weather conditions and under predicted extreme load and weather conditions;
  (iii) a simple drawing of a cross section of the proposed right-of-way of the HV line and any adjoining right-of-ways showing the placement of the supporting structures at typical locations, with the height and width of the structures, the width of the right-of-way, and the lateral distance between the conductors and the edge of the right-of-way indicated; and
  (iv) a system map which shows a suitable detail the location and voltage of all existing transmission lines and substations of the applicant and the location and voltage of the proposed HV line and associated substations;
(14) a statement identifying any litigation concluded or in progress which concerns any property or matter relating to the proposed HV line, right-of-way route, or environmental matters; and
(15) such additional information as the Commission may require

Source: Commonwealth of Pennsylvania 52 Pennsylvania Code Subchapter G, §857.71 through §57.57.

megawatts in winter and 21,141 megawatts in summer for the seven largest investor-owned electric utilities in the state. These companies had 28,365 megawatts of capacity as of December 31, 1985 (Pennsylvania Public Utility Commission 1986, 11). It is estimated that by 1994 generating resources will be 29,254 megawatts in winter and 28,328 megawatts in summer (Pennsylvania Public Utility Commission 1985, 18). Pennsylvania utilities share of the new capacity to come on-line by the year 2000 is estimated to be over 4,400 megawatts. Some of this planned capacity may not come forth if nuclear power plants are abandoned.

However, there does not appear to be a crucial need for new services in Pennsylvania. There is increasing stress on the transmission system by the addition of cogenerators and small power producers. Pennsylvania Power and Light has over 400 megawatts of nonutility-generated power (Pennsylvania Public Utility Commission 1986, 12). If new resources are added, new transmission lines may be needed to sell power beyond the local area.

Even though Pennsylvania residents do not face an imminent power supply crisis, the construction of new transmission lines can ease congestion in the system and improve the ability of small power producers in Pennsylvania to market their power. Some of this power will be generated at culm stations, and Pennsylvania's Department of Natural Resources encourages these projects to reduce the problem of culm piles throughout the state. The construction of new high-voltage transmission lines designed to serve customers in other states may have the direct benefit of improving transfers of power from these small power producers. In this light, a large transmission project may be favorably viewed by the public utility commission.

PUBLIC ACCEPTANCE OF NEW ENERGY FACILITIES*

Siting energy facilities is often a problem because the project meets with public opposition. Opposition to energy projects has occurred throughout the United States and abroad for many types of energy projects (also nonenergy-related projects). There has been public opposition to power plants (hydroelectric, nuclear, and coal fired), transmission lines, terminals, and other conversion facilities. The opposition has come from local citizens' groups, public interest groups (environmental and consumer related), and individual citizens.

Opposition to nuclear power is much more widespread than is opposition to coal-fired power plants. Indeed, Paul Slovic (1987) has found that individuals fear nuclear power plants much more so than conventional power plants both because of the dread risk of nuclear facilities and the unknown nature of the risk of nuclear power. Certain interest groups, such as the Sierra Club, have anti-nuclear positions and will fight nuclear proposals irrespective of the location selected. For coal-fired power plants, the extent of public opposition depends more upon the location selected for the facility. Particularly sensitive environments that have been selected for coal-fired power plants have met with strong opposition. An example was the Kaiparowits project in southern Utah. Opposition also depends upon the benefits a local community will enjoy, compared to the sacrifices it must face.

Industrial development of all types produces both positive and negative impacts for local communities

---

*Greg Sayre assisted in the preparation of this section.

(Rogers et al. 1978; Summers 1977; Summers et al. 1976). These range from increasing the cost of public services (Summers et al. 1976), generating new jobs (Bertrand and Osborne 1959), reversed out-migration (Andrews and Baunder 1967; Summers 1973), and increases in local expenditures (Brinkman 1973; Garrison 1970; Molotch 1976).
The benefits and costs of industrial development are also not equally distributed among the individuals within the impacted area (Molotch 1976; Rogers et al. 1978). While some segments of the rural population will benefit more from new developments than others, rural communities generally have favorable attitudes toward economic development (Murdock and Leistritz 1979; Bachtel and Molnar 1981; Stout-Wiegand and Trent 1983; Rogers et al. 1978; Summers and Clemente 1976). In fact, opposition to development in rural areas has been shown to be insignificant by many researchers (Scott and Summers 1974; Summers 1973; Summers 1977; Molotch 1976; Trent, Stout-Wiegand, and Smith 1986).
Public opposition to coal-fired power plants depends upon whether they are located in an environmentally sensitive area and whether the local or regional economy is likely to receive great economic impacts from the facility. Such facilities are more likely to stir public controversy if the facility conflicts with the local environment and if the economic benefits to the local or regional economy are minimal.
Based upon recent experience in West Virginia and the results of elite and public opinion surveys, it is not likely that proposals to site new coal-fired power plants in West Virginia will cause much public or citizen group opposition. Such opposition to coal-fired power plants seem to be more likely to occur in areas outside the coalfields, even if careful site planning is used to select locations for new power plants to minimize impacts. Some recent experiences in siting coal-fired power plants in the Appalachians are described in the following subsection, and the results of surveys of environmental interest group leaders and inhabitants near existing power plants are then summarized to ascertain their feelings toward these facilities and their willingness to see new facilities constructed in their area. Finally, these results are compared with surveys of people who live in areas of West Virginia and Pennsylvania that do not have nearby power plants.

The Experience in Pinesburg, Maryland

In 1971 the state of Maryland established the Maryland Power Plant Siting Program to identify and purchase sites for new power plants that were found to be acceptable across a range of economic, engineering, environmental, and social criteria. This was accomplished through the use of the Maryland Automated Geographic Information (MAGI) data, base which contains information on physical, cultural, administrative, and other conditions for 92-acre cells throughout parts of Maryland involved in particular studies. A site suitability reening procedure is followed to identify candidate areas and, finally, candidate sites for new power plants that are acceptable across a range of criteria (Calzonetti, Sayre, and Spooner 1987; Dobson 1979). The program was instituted for the following reasons:

a. To make certain that sites for power plants that meet the necessary engineering, environmental, and social siting criteria are available when needed in the future.

b. To facilitate public and local governmental participation in power plant siting decisions.

c. To air controversial issues regarding power plant siting well before there is a critical need for the new plant so that delay in the siting process can be reduced.

d. To provide alternatives to utility-proposed sites, thus making it possible for the state to deny a request by a utility to site a power plant in an unacceptable site given the availability of an acceptable state-owned site.

It was maintained that this program would lead to the selection of better sites and that projects could actually be brought on-line much more quickly without prolonged debate and public outcry.

In 1979 the Western Maryland Power Plant Study was undertaken in response to a mandate by the Maryland General Assembly to purchase at least one site for use by the Potomac Edison Company in the region. After long study and public meetings, it was decided to bank a site in Pinesburg, near Hagerstown, for a 1,890-megawatt coal-fired power plant. The Potomac Edison Company had no intentions of siting a power plant at that location at that time, but in fact the parent company, Allegheny Power, was intending to locate its next plant in Pennsylvania. However, the state was proceeding to acquire this land for use in the future for a coal-fired power plant.

The state had contact with local representatives and officials and felt that there was strong support for locating the projected power plant in Pinesburg. The prospect of $28 million in annual tax revenues and new jobs was attractive to many local officials. It was also believed that the plant would not have any noticeable impacts.

As plans to acquire land at Pinesburg continued, residents on and adjacent to the site became aware of the intentions of the power plant siting program. It happened that a veterinarian was building a home adjacent to the site, and a physician owned a home that was to be condemned by the state. These individuals were instrumental in organizing an ad hoc citizens' group, Citizens for the Protection of Washington County (CPWC).

These individuals educated themselves about power plant operations and found individuals who lived near other power plants to identify complaints. Because the Maryland Power Plant Siting Program is within the state's Department of Natural Resources, a state agency, the program provided all the documentation and information to the interest group requested. Much of this information was later used against the power plant proposal. Potomac Edison also provided information to the citizen's group. It is not clear whether such information would have been forthcoming if the proposal had come from a private utility.

The interest group broadened the issue of the new power plant to involve the greater Hagerstown community in order to defeat the proposal. One of the main arguments against the new plant concerned the health impacts of the proposed plant. "Fact Sheets" were distributed at many doctors' offices informing people that they might suffer respiratory and allergy problems if the plant came into operation. The CPWC also called attention to the aesthetic impacts and rejected the power plant siting program's claim that the new facility would be a boost to the local economy. The interest group maintained that the stacks and cooling towers would dominate the valley and that the new transmission lines would be unsightly. The CPWC also noted that the new plant was not needed. Electricity demand projections were down, and there was no need to construct a new power plant at that time.

Some local officials supported the plant in the early stages of the process because of the promised tax revenues and employment opportunities. The CPWC said that, on the contrary, the area would lose jobs and property values would decline. They argued that a new power plant would produce additional air emission and would preclude the siting of other industry which might offer more jobs and greater local multipliers. The interest group also said that the plant would employ only a few more than 100 workers and that most of the jobs from the new plant would occur in West Virginia and Pennsylvania where the coal for the plant was to be mined. They said that the plant should be built in West Virginia because the people in that state would like to have the plant. The CPWC also said that some of the industry that had moved to Hagerstown because of its aesthetic qualities would relocate if the plant was built. Finally, the CPWC said that the plant would cause property values to decline as the town would be less attractive and more polluted, and would be losing industry.

These claims were really never attacked by the Maryland power plant siting program or by Potomac Edison. The Power Plant Siting Program's administrator was instructed not to engage in public debate over the plant and to avoid controversy. If the plant had been proposed by a private utility, there would have been a campaign to discredit many of the arguments used by the CPWC.

By the fall of 1985, 24,000 signatures were obtained in opposition to the site acquisition program. A contingent of Hagerstown politicians and CPWC representatives visited the governor of Maryland, who later announced a moratorium on the site acquisition program. In the winter of 1986, delegates from western Maryland introduced bills in the Maryland legislative session to modify the power plant siting act by removing the requirement that the Department of Natural Resources acquire a power plant site. The bill was signed into law, and no further action at the Pinesburg site has been taken.

In this case, a coal-fired power plant was defeated because of the action of an ad hoc citizens' group which arose from the discontent of a few individuals who owned property on or adjacent to the proposed plant site. It really is not possible to predict whether a proposal to build any facility will lead to opposition of this type from individuals whose area is to be directly impacted by the development. Condemnations will always affect certain individuals, and some people will resist the loss of their property. It generally takes special individuals, or the activities of larger interest groups, to expand personal losses from such condemnations into wider community resistance.

It was easy for the CPWC to maintain that the local community would not gain very much from this new plant and to raise fears about the impacts of this facility. Would it be possible to do this in an area where the people are more familiar with modern coal-fired power plants and where the new power plant would lead to long-term markets for the regional coal industry? In Hagerstown there was not that much familiarity with modern coal-fired power plants. In fact, the power plant siting program took advisory board and CPWC members on tours of modern coal-fired power plants in West Virginia in order to explain how these facilities operate. The CPWC would have been less successful if the local community residents believed that the power plant would be able to improve economic conditions in the area.

In West Virginia there is widespread support for projects that lead to better markets for the coal industry. In some cases, however, proposed facilities have met with public opposition. For example, a coal liquefaction plant was proposed to be constructed near Morgantown, West Virginia, in 1979, but it met with strong local opposition. This plant would be more similar to a chemical plant than to a power plant, and Morgantown is the site of West Virginia University which had many faculty and students who were not enthusiastic about the plant. There is no record of the public resisting the construction of coal-fired power plants that have been constructed in the state. An example of the nature of such community response can be seen in selected West Virginia power plant proposals.

Reactions to West Virginia Power Plant Projects

On January 24, 1974, a joint news conference was held by West Virginia Governor Arch A. Moore, Jr., and Donald C. Cook, chairman of the American Electric Power Company, announcing the decision to construct a 1,300-megawatt coal-fired power plant near New Haven, West Virginia. This plant was later named the Mountaineer power plant. It was stated that this plant would burn 3.8 million tons of West

Virginia coal annually. Cook explained that the site was chosen because of its location on the Ohio River, its proximity to West Virginia coal, its accessibility to the transmission network, and the area's "cooperative labor climate." The governor called attention to the economic benefits of the new plant. In addition to improved coal markets, the plant would employ up to 2,600 construction workers for the three-and-a-half-year construction period, a construction payroll of over $135 million, permanent employment for 150 workers, a permanent payroll of $1.35 million per year, and millions of dollars in local taxes and purchases of local goods and services (Appalachian Power Company 1974).

On February 7, 1974, Appalachian Power Company (the operating subsidiary of American Electric Power) held a meeting and answered questions from local residents about the new plant. Some concerns were raised about the possible environmental impacts of the plant. At the meeting, the mayor of Point Pleasant said, "We are extremely pleased to have the plant here. . . . if there is any way the city can help, just let us know" (Baisden 1974). At this meeting, Appalachian Power Company said that the plant should go on-line in 1977. By June, however, American Electric Power announced that the construction schedule would be slowed down in an economy move, thus the completion date would be pushed back from 1977. The plant began generating power on September 15, 1980, six-and-one-half years after the announcement was made. This can be considered a successful siting project, going from the initial announcement to completion and operation with minimal delay and opposition.

The announcements for the construction of the John E. Amos plant, and an additional unit that was added later, were also met with enthusiasm; this facility was completed without delay and with public support. On June 20, 1968, Appalachian Power Company announced that it would build a 1,600-megawatt coal-fired power plant in Putnam County, West Virginia, named the John E. Amos plant. The local newspaper portrayed the local enthusiasm for the new project. According to the <u>Putnam Democrat</u>, ". . . Announcement of Appalachian Power Company's mammoth new generating plant in Putnam County was greeted with cheers and rejoicing. Probably the greatest single event in Putnam's history." According to the company, one reason for the selection of the site was the high degree of cooperation provided by the local residents. Even before the first two units were completed, American Electric Power decided to add another unit at the plant, increasing its size to 2,900 megawatts, which was the largest in the country at that time. The first 800-megawatt unit was scheduled for completion in July, 1971; the second 800-megawatt unit was to go on-line in March 1972; and the third unit, rated at 1,300 megawatts was to begin commercial operation in October 1973. It was acknowledged when the additional unit was announced that much of the power would be sold outside West Virginia. These units went into service as scheduled without any public resistance.

Similar stories could be told for all of the other coal-fired power plants in West Virginia. By and large, the plants are supported by both the local community and state government. Also, there is support from other strong groups in the state, namely the labor unions, the coal miners, and their representatives in the state legislature. This does not mean that opposition to a new plant could not occur in the future, but there is an excellent track record for having power plants completed on schedule with much public support. In the next section, we report on surveys conducted to gauge the likely support or opposition to new power plants planned for West Virginia. We first summarize the results of a survey of environmental interest group leaders, then we turn to public opinion surveys of residents in communities with and without nearby coal-fired power plants.

SURVEY OF WEST VIRGINIA ENVIRONMENTAL INTEREST GROUP LEADERS

In the summer and early fall of 1984, before Governor Moore announced that he would be building state-owned power plants, the group leaders of leading environmental interest groups in West Virginia were interviewed. The purpose of this survey was to identify the interest group leaders' views toward various energy technologies, to determine whether they would support the construction of energy facilities in various parts of the state, and to ascertain whether they had feelings about the state's electricity export role. Organizations contacted included the environmental issue groups, naturalist organizations, energy-consumer organizations, and health organizations that had previous dealings in environmental issues. Nineteen surveys were conducted (Calzonetti 1985).

Three-fourths of the respondents favored West Virginia's role as an energy-exporting state, and two-thirds of the respondents were in favor of exporting electricity to other states. However, the respondents pointed out that this must be done in an environmentally sound manner. Forty percent of the respondents believed that construction of new coal-fired capacity in West Virginia would result in some economic benefit to the state. The respondents almost entirely opposed nuclear power as an electricity-generating source. Over two-thirds of the interest group leaders believed that coal combustion in power plants can be conducted in an environmentally acceptable manner in West Virginia; however, 18 percent believed that coal combustion is never conducted in an environmentally sound manner. Their complaint was that not enough was being done to combat the acid rain problem, which they felt to be the most important environmental problem facing the state of West Virginia (followed by hazardous waste disposal).

The respondents were also asked for their opinions of new energy facilities being sited in various places in the state. For the most part, there would be opposition from these individuals if energy facilities of any type were located in the Appalachian highlands areas. This region has many parks and natural areas that these individuals wish to protect from development or environmental disruption.

In sum, the results of this survey suggest that environmental groups will not be quick to attack new coal-fired power plants, built with scrubbers or using fluidized-bed technology, so long as they are not located in the Appalachian highlands. These individuals recognize the positive economic impacts of these facilities and believe that the facilities can be operated in an environmentally sensitive manner. The precise location for such a facility may influence environmental group reactions. For instance, a location near a wetland or historical area may cause some concern that could lead to opposition.

PUBLIC OPINION SURVEYS**

In the Pinesburg, Maryland, case it was not an environmental interest group that instigated the opposition to the proposed power plant, but landowners who were able to expand the issue. As noted earlier, it is impossible to predict this type of opposition. However, one can gain a general sentiment toward existing and hypothetical facilities through the use of public opinion surveys.

Public opinion surveys were conducted by the authors, during the summer of 1986, in five communities: four in West Virginia and one in Pennsylvania. Three of the West Virginia communities had nearby coal-fired power plants: Belmont, Mt. Storm, and New Haven. Residents in the community of Elizabeth, West Virginia, which does not have a power plant, were also surveyed, as were residents of Honesdale, Pennsylvania, a community also without a power plant in the northeastern part of the state. These two communities were selected because of their economic similarity to the West Virginia communities with power plants. The purpose of the survey was to identify the reported impacts of power plants in those communities with these facilities, and to measure the level of support or opposition to the location of additional power plants near the communities.

In contrast to the key informant survey conducted by Stephen Webb, R. Krannich, and F. Clemente (1980), this study utilized a ten-percent random sample of telephone listings from local telephone directories. The use of telephone listings is not optimal, but its listings are considered better sources in rural areas than in urban areas where many people have unlisted numbers. Table 44 lists the response rate, total number of telephone calls made, and the characteristics of the respondents in each of the five communities. For each community surveyed, questions were asked about the perceived quality of life and the existing economic and environmental conditions in the area. Table 45 summarizes the responses to a question asking the respondents to appraise the outlook for their

---

**Greg Sayre assisted in the preparation of this section.

community. Only the community of New Haven reported a dismal outlook. Fifty percent of the respondents said that they thought that the community was declining, probably in response to the loss of a few manufacturing plants in the area. The respondents in the other communities held positive feelings about their prospects for the community.

The respondents in Belmont, Mt. Storm, and New Haven were asked if they supported or opposed the construction of the power plant when it was planned. Table 46 shows that very few of the residents in these communities opposed the idea of a new coal-fired power plant in their community. Sixty-five percent of the respondents in Belmont, 39 percent of those in Mt. Storm, and 81 percent of those in New Haven were in favor of the construction of the plant. The community of Mt. Storm showed the least initial support as 47 percent of the respondents reported that they were indifferent to the plant when it was initially proposed. There was some opposition to the proposed plants, even though it was slight. Opposition was greatest in the Mt. Storm area. This area did not have very much coal mining activity until the Mt. Storm plant was built.

Residents in the communities with power plants were asked about the undesirable impacts of these facilities. Table 47 shows that only 28 percent of the respondents reported undesirable impacts from the power plant operations. Almost 70 percent of those surveyed said that they did not experience any undesirable impacts from the power plants. More of the people in the vicinity of the Mt. Storm power plant reported undesirable impacts from the plants than did those people living near the Pleasants and Mountaineer plants. The plant at Mt. Storm is located several miles from the community and is hidden by the mountainous terrain in the area. In the other two locations, the plants are in the Ohio Valley and are visible to most of the people living in the valley.

The respondents were also asked about how they would react to the construction of a new power plant unit in their community (Table 48). Respondents from the West Virginia communities with power plants overwhelmingly supported the development of additional generating capacity in their areas. In each case, over 80 percent of those questioned favored such developments. On average, only 8 percent of those interviewed in these three communities reported being opposed to additional facilities. People in the two communities without power plants voiced less support for the construction of new power plants. In Elizabeth, West Virginia, 56 percent favored the idea and 30 opposed; in Honesdale, 35 percent favored, and 43 percent opposed. Also, few people in communities with power plants were indifferent-most had an opinion. On the other hand, many of those questioned in Elizabeth and Honesdale claimed to be indifferent to such developments. It was obvious from our interviews that many of those in Elizabeth and Honesdale did not know enough about power plants to formulate an opinion. There appeared to be more ignorance in Honesdale than in Elizabeth. Some in Honesdale were even unable to distinguish between coal-fired and nuclear power plants.

Table 44
Response Rate

Community	Number	Sample Size (%)	Male (%)	Female (%)
Belmont	109	10	40	60
Mt. Storm	60	10	42	58
New Haven	38	20	37	63
Elizabeth	60	10	32	68
Honesdale	110	5	25	75

Note: Analysis based on a 45-percent response rate, including unanswered calls, disconnected numbers, and refusals.

Source: Data collected by the authors.

Table 45
Attitudes Toward the Present Condition of the Community (percentage of total respondents)

Community	Improving	Staying the Same	Declining	Don't Know
Belmont	63	28	7	0
Mt. Storm	50	32	16	4
New Haven	15	32	50	3
Elizabeth	54	27	17	2
Honesdale	57	34	7	2

Source: Data collected by the authors. Totals may not equal 100 because of rounding.

Table 46
Attitudes toward the Construction of Recent Power Plants
(percentage of total respondents)

Community	Strongly Favored	Favored	Neither Favored nor Opposed	Opposed	Strongly Opposed	Don't Know
Belmont	18	47	15	7	0	13
Mt. Storm	28	11	47	8	3	3
New Haven	42	39	6	5	1	7

Source: Data collected by the authors. Totals may not equal 100 because of rounding.

Table 47
Undesirable Impacts of Power Plants (percentage of total respondents)

Community	Yes	No	Don't Know
Belmont	27	68	5
Mt. Storm	37	63	0
New Haven	20	71	8

Source: Data collected by the authors. Totals may not equal 100 because of rounding.

Following up on attitudes toward the general development of coal-fired power plants in the communities, respondents were asked what the minimum distance the plant would have to be from their home before it would be acceptable to them. Citizens from Belmont and New Haven indicated that a new facility would be acceptable to them at much closer distances to their homes than those from Mt. Storm, Elizabeth, and Honesdale. While the majority of Belmont and New Haven residents indicated that a facility less than five miles away would be acceptable, residents of Mt. Storm, Elizabeth, and Honesdale generally wanted the plant to be built more than five miles from their home. In addition, a substantial number of residents from Belmont and New Haven considered that less than a mile and from one to two miles would be minimum acceptable distances from their homes.

In comparison to coal-burning facilities, nuclear power plants were perceived much more negatively as indicated in Table 49. Even in New Haven, where recent worker layoffs would seem to put a premium on any new job creation, only 18 percent of those interviewed favored such a facility.

CONCLUSION

This analysis allows us to draw a number of conclusions. Appalachian communities with power plants tend to have more favorable attitudes toward the development of additional facilities than communities without power plants. This could be several reasons for this discrepancy. Perhaps those individuals who did not wish to live near a power plant vacated the area when the power plant was built so that those remaining were those initially in favor of the plant. It is also possible that people found out that the new plant did not produce many serious impacts, or if so, they learned to adjust to these. The impacts listed that annoyed some people included the noise of the facility (sometimes even from the loudspeaker), dust, and traffic. Even in those communities where the attitudes were favorable, people preferred them at a greater distance from their homes.

Table 48
Attitudes Toward the Construction of a Hypothetical Coal-Fired Power Plant
(percentage of total respondents)

Community	Strongly Favor	Favor	Neither Favor nor Oppose	Oppose	Strongly Oppose	Don't Know
Belmont	12	72	2	7	2	7
Mount Storm	19	65	5	3	5	3
New Haven	29	56	4	6	3	3
Elizabeth	21	35	14	21	9	2
Honesdale	6	29	20	31	12	1

Source: Data collected by the authors. Totals may not equal 100 because of rounding.

Table 49
Attitudes Toward the Construction of a Hypothetical Nuclear Power Plant
(percentage of total respondents)

Community	Strongly Favor	Favor	Neither Favor nor Oppose	Oppose	Strongly Oppose	Don't Know
Belmont	3	10	3	22	58	3
Mount Storm	0	17	19	11	39	14
New Haven	4	14	7	14	57	7
Elizabeth	2	9	14	9	57	10
Honesdale	3	6	11		66	3

Source: Data collected by the authors. Totals may not equal 100 because of rounding.

While opposition to current coal-fired power plants has taken place in other areas, such as Pinesburg, it seems that in West Virginia at least substantial portions of the population would welcome additional power plant construction activity. Power plants are viewed by many in communities with power plants to be good neighbors and are seen to benefit the community through taxes and jobs. Even though parts of West Virginia are in severe decline, the strong opposition to nuclear power indicates that people are selective about the types of developments they would welcome in their community. Thus, the support for new power plant construction does not mean that these individuals would accept any type of development just to gain jobs and tax revenues.

BIBLIOGRAPHY

Andrews, W. H., and W. W. Baunder. 1967. The Effects of Industrialization on Rural Counties: Comparisons of Social Changes in Monroe and Noble Counties of Ohio. Department Series A, E, 407. Wooster, Ohio: Ohio Agricultural Research and Development Center.

Appalachian Power Company. January 25, 1974. Press release.

Bachtel, D. C., and J. J. Molnar. 1981. "Women As Community Decision Makers." Human Services in the Rural Environment 6; 2:3-10.

Baisden, Harry L. February 8, 1974. "Power Plant Plans Clean Air Devices." Huntington Herald Dispatch, 13.

Bertrand, Alvin L. and Harold W. Osborne. 1959. "The Impact of Industrialization on a Rural Community." Journal of Farm Economics 41:1127-34.

Brinkman, George. 1973. "Effects of Industrializing Small Communities." Journal of the Community Development Society 4:69-80.

Calzonetti, F. 1985. Survey of West Virginia Environmental Group Leaders. Morgantown, W.Va.: West Virginia University Energy and Water Research Center Report.

Calzonetti, F., Gregory G. Sayre, and Derek Spooner. 1987. "A Reassessment of Site Suitability Analysis for Power Plant Siting: A Maryland Example." Applied Geography 7:223-41.

Dobson, J. E. 1079. "A Regional Screening Procedure for Land Use Suitability Analysis." The Geographical Review 69:224-34.

ECAR/MAAC Coordinating Group. 1985. "ECAR/MAAC Interregional Power Transfer Analysis." presentation to the U.S. Department of Energy. Washington, D.C. (June).

Garrison, Charles B. 1970. The Impact of New Industry on Local Government Finances in Five Small Towns in Kentucky. Washington, D.C.: U.S. Department of Agriculture, Economic Research Service, Agricultural Economic Report no. 141.

Molotch, H. 1976. "The City As a Growth Machine: Toward a Political Economy and Place." American Journal of Sociology 82:309-32.

Morgan, M. Granger, H. Keith Florig, Indira Nair, and Gordon L. Hester. 1988. "Controlling Exposure to Transmission Line Electromagnetic Fields: A Regulatory Approach That Is Compatible with Available Science" Public Utilities Fortnightly 121, 6 (March 17):49-58.

Morgan, M. Granger, H. Keigh Florig, Indira Nair, and D. Lincoln. 1985. "Power-Line Fields and Human Health." IEEE Spectrum (February):62-68.

Murdock, S. H., and F. L. Leistritz. 1979. Energy Development in the Western United States: Impacts on Rural Areas. New York: Praeger Press.

Myhra, D. 1977. "Fossil Projects Need Siting Help Too," Public Utilities Fortnightly 100 (September):24-28.

National Governors' Association. 1987. Moving Power: Flexibility for the Future. Washington, D.C.: National Governors' Association.

Rock, J. M. 1977 "No Boomtown on the Kaiparowits Plateau: Who Made the Decision and Why?" *Intellect* 105 (February):248-50.

Rogers, D. L., B. F. Pendleton, W. L. Goudy, and R. O. Richards. 1978. "Industrialization, Income Benefits, and the Rural Community." *Rural Sociology* 43:250-264.

Slovic, Paul. 1987. "Perceptions of Risk." *Science* 236, (April):280-85.

Stout-Wiegand, N. Trent, and R. B. Trent. 1983. "Sex Differences in Attitudes toward New Energy Resource Developments." *Rural Sociology* 48:637-46.

Summers, G. F. 1977. "Industrial Development in Rural America: A Quarter Century of Experience." *Journal of the Community Development Society* 8:6-18.

---. 1973. *Large Industry in Rural Areas: Demographic, Economic, and Social Impacts*. Working paper RID 73.19. Madison: University of Wisconsin. Center for Applied Sociology.

Summers, G. F., and F. Clemente. 1976. "Industrial Development Income, Distribution, and Public Policy." *Rural Sociology* 41:248-68.

Summers, G. F., et al. 1976. *Industrial Invasion of Non-Metropolitan America: Quarter Century of Experience*. New York: Praeger Press.

Trent, Roger B., N. Stout-Wiegand, and D. K. Smith. 1986. "Attitudes towards New Development in Three Appalachian Counties." *Growth and Change* 16:70-86.

U.S. Army Corps of Engineers. 1977. *Final Environmental Impact Statement, Project 1301, New Power Plant on the Ohio River*. Huntington, W.Va.: U.S. Army Engineer District.

Webb, Stephen, R. Krannich, and F. Clemente. 1980. "Power Plants in Rural Area Communities: Their Size, Type and Perceived Impacts." *Journal of the Community Development Society* 11:81-95.

9

A Cost Comparison: Mine-Mouth versus Load Center Power Plant Locations

Frank J. Calzonetti, Timothy Allison, Muhammad A. Choudhry, and Tom Torries

One approach to use in examining the feasibility of coal-by-wire is to compare the cost of constructing a new base load power plant in the coalfields and transmitting the power to markets to the cost of shipping coal to a power plant located near the load center. This chapter compares the cost of building a new base load, pulverized, coal-fired power plant in West Virginia and transmitting the power to the East to building equivalent coal-fired power plants in New Jersey and Connecticut and using coal transported from West Virginia. This type of analysis, although interesting, is not necessarily the approach that would be taken by power company planners interested in their future electricity supplies. A comparison of equivalent power plants across locations would be conducted after other electricity supply, conservation, and load management alternatives were considered and the type of power plant had been selected. However, the feasibility of future expansion of coal-by-wire would be dismal if a comparison of delivered costs indicated that coalfield power plants did not enjoy economies over load center facilities. The chapter is limited purely to the comparative costs of electricity production; it does not examine in detail differential environmental and political considerations.

The chapter is divided into four sections. The first section summarizes other studies that have evaluated coal-by-wire options. The second section describes a variety of models used to calculate the costs of generating electricity at new coal-fired power plants. The third section reports the results of scenarios that compare coal-by-wire to the construction of new power plants at eastern load centers. A rationale for the comparative cost scenarios chosen is given, followed by a description of the assumptions used in the costing model. The results of the power costing comparisons are then described. Three plant sizes (300 megawatts, 500 megawatts, and 750 megawatts) are used with three construction schedule scenarios. Finally, conclusions based on the results of the costing analysis are offered.

RELATED STUDIES

Two recent studies have evaluated coal-by-wire concepts to the Northeast: C. Neme's <u>Midwest Coal by Wire</u> for the Center for Clean Air Policy and A. F. Destribats' "Power From the Midwest" for New England Electric System. The objective of the study completed by the Center for Clean Air Policy was to determine whether it was possible to add scrubbers to existing power plants in the Midwest (Ohio), to build new transmission facilities, and to deliver power more cheaply than by means of the other power supply alternatives open to the region. New England Electric System conducted an in-house study to compare coal-by-wire to other alternatives. Both studies conclude that coal-by-wire using existing power plants is less expensive than building new power plants in New England.

Center for Clean Air Policy's <u>Midwest Coal by Wire</u>

<u>Midwest Coal by Wire</u> evaluates two configurations for supplying the Northeast with midwestern power. The first involves the sale of 1,000 megawatts of firm power from American Electric Power's (AEP's) Gavin plant to the New England Power Pool (NEPOOL). The second scheme provides for a sale of 2,000 megawatts of power from the Midwest: 1,000 megawatts to be purchased by the Pennsylvania-New Jersey-Maryland (PJM) interconnection and 1,000 megawatts to be purchased by NEPOOL. The Gavin plant was selected because it is one of the major producers of sulfur dioxide in the country and it enjoys the cheapest bus bar costs on the AEP system.

Table 50 summarizes the results of the first configuration. The costs of power delivered to New England are a sum of the energy costs, the scrubber capital costs, and the power line costs, all of which are borne by NEPOOL. The Gavin plant is to provide "surplus" energy to New England until the year 2000, when additional capacity will need to be completed (either constructed by utilities or the state of West Virginia). According to this study, the total cost of power ranges from between 4.83 to 5.62 cents/kilowatt-hour in 1995. This is exceedingly low compared to any other alternative except Canadian hydroelectric power. The study compares the costs of coal-by-wire to the cost of NEPOOL oil and gas at the time to compute the savings to NEPOOL, which would be very substantial in the twenty-first century.

The study makes some heroic assumptions that have been criticized by the utility industry and by a review of the study by the AEP. For example, NEPOOL pays for the installation of a scrubber on the Gavin plant as well as for the construction of a ± 450-kilovolt high-voltage direct current (DC) transmission line from West Virginia to Connecticut (540 miles in length). The major shortcomings of the report are the following: (1) no capacity charge is provided for the firm AEP power which is provided to the year 2000; (2) the study does not compute the net capacity penalty of the scrubber or the operation and maintenance of the scrubber; and (3) transmission losses on the new

Table 50
Midwest Coal-by-Wire Costs of Energy, Scrubbers, and New Capacity (cents/kWh)

	1995	2000	2010
Energy Costs	2.45	2.15	2.09
Scrubber Capital Cost	0.94	0.94	0.94
New Plant Capital Cost	0	2.79	2.79
TOTAL PAYMENTS TO MIDWEST	3.4	5.87	5.82
POWER LINE COSTS (1000 MW Sale to NEPOOL)	1.43-2.42	1.43-2.42	1.43-2.42
TOTAL COST FOR POWER	4.83-5.62	7.30-8.29	7.25-8.24

Notes: Figures are from ICF, Inc. analysis of the costs of acid rain controls to AEP (conducted for the Center for Clean Air Policy), inflated to 1987 dollars.

Energy costs include both fuel costs and operation and maintenance costs.

Capital costs for scrubbers and new capacity (1200 MW) are levelized over 30 years. No capital charge is assumed until NEPOOL purchase requires that new capacity be built (the year 2000).

Power line cost estimates are for configuration 1 of the power line where the midwest would simply sell 1000 MW of firm power to the Northeast. Under configuration 2, where Pennsylvania-New Jersey-New York share the costs of the line, the costs to NEPOOL would be lower (1.08-1.69 cents/kWh). They include the costs for converter stations but do not include costs for AC reinforcements in the NEPOOL system because they can only be determined through detailed computer analysis of system reliability. They will likely be less than 10% of the power line costs. The low cost estimate is based on the New England Hydro-Transmission Corporation application to build a power line from Monroe New Hampshire to Sandy Pond, Massachusetts; the high cost estimate is based on two "conceptual plans" for Midwest-Northeast transmission lines presented in a 1986 NEPOOL report.

Source: C. Neme, Midwest Coal by Wire, 1987.

transmission line, which should increase the delivered energy costs by about 5 percent, have been ignored.

It is doubtful whether the Gavin plant could be dedicated to provide the 1,000-megawatt sale to New England. This would take the plant away from the AEP and raise rates for AEP customers because the cheapest plant would not be available to serve local needs, and other, more expensive units would need to be used to serve peak loads. Table 51 provides AEP's revision of the Center for Clean Air's analysis which includes the neglected cost factors (and other assumptions) for 1995. The result is a total cost of energy as received in New England ranging from between 9.06 to 10.11 cents/kilowatt-hour (1987

Table 51
Evaluation of Midwestern Power (Firm) to the Northeast
(Comparison of AEP versus Center for Clean Air Policy Estimates, in cents/kWh)

	1995 CCAP Estimate	1995 AEP Estimate	2000 CCAP Estimate	2000 AEP Estimate
Energy Cost	2.45	2.11	2.15	2.11
Demand Cost	0	2.34	0	0
Scrubber Cost				
Capital	0.94	1.32	0.94	1.32
O & M	0	1.08	0	1.08
Scrubber Replacement Capacity				
Capital	0	0	0	0.54
O & M	0	0	0	0.04
New Plant Capacity				
Capital	0	0	2.79	3.83
O & M	0	0	0	0.3
Generation Subtotal	3.39	7.18	5.88	9.22
Power Line Cost	1.43-2.42	1.43-2.42	1.43-2.42	1.43-2.42
TOTAL COST OF ENERGY AS DELIVERED	4.82-5.81	8.61-9.60	7.31-8.30	10.65-11.64
TOTAL COST OF ENERGY AS RECEIVED IN NEW ENGLAND	------	9.06-10.11	------	11.21-12.25

Notes for AEP Estimates:
   All calculations assume 1000 MW of firm power sale, 100% load factor capacity cost based on 1200 MW of capacity dedicated to the sale

   For the Basis for Calculation of AEP Estimates, see Appendix 2.

Source: American Electric Power Service Corporation, Critique of Center for Clean Air Policy Report, 1988.

dollars). Using these results, coal-by-wire is far less attractive than other local alternatives.

New England Electric System's "Power From the Midwest"

At the New England Governor's Conference Northeast-Midwest Transmission Workshop held in March 1988, Alan Destribats of New England Power Service Company presented the results of an internal study on coal-by-wire (Destribats 1988). The results of this study are more comparable to AEP's results than to those provided by the Center for Clean Air Policy. The concept evaluated was a purchase of 2,000-megawatts from West Virginia and western Pennsylvania: 1,000 megawatts to be provided to New York and 1,000 megawatts to be sold to New England. A comparison of alternating current (AC) and DC lines led New England Electric to conclude that a three-terminal 500-kilovolt DC scheme would be less costly. This would include a western terminal rated at 2,000 megawatts and terminals in New York and New England rated at 1,500 megawatts each. The total cost of the line would be about $1 billion.

Table 52 summarizes the cost in 1996 for an existing (which does include the capacity cost of the existing plant) and a new power plant in addition to the DC line. For 1996, the total cost for using an existing midwestern plant would be 10.6 cents/kilowatt-hour, which is very close to the upper limit of the AEP study. If a new plant were constructed in the Midwest, the cost would be 18.8 cents/kilowatt-hour.

Table 53 compares the thirty-year levelized costs to other alternatives. In 1987 dollars, the cost of power from an existing midwestern plant with the construction of the new line would be 5.3 cents/kilowatt-hour. Power from Hydro Quebec is shown to be less expensive than this, but no other alternative can beat coal-by-wire. In summary, the study suggests that obtaining power from existing power plants in the coalfields is less expensive than constructing new generating facilities in New England; however, if new power plants are built, it would be cheaper to have the facilities located in New England than in the coalfields.

West Virginia University's Study

The above discussion indicates that there is some discrepancy between the projection of the cost of providing power to New England from the Midwest. The remainder of this chapter is devoted to the description and discussion of an independent evaluation of the economics of a coal-by-wire concept to New Jersey and New England. This evaluation is based upon the Consolidation Coal Company Utility Power Costing Model and West Virginia University's (WVU's) electricity transmission model. These results are closer to those depicted by the AEP and New England Electric than to those provided by the Center for Clean Air Policy.

ESTIMATING POWER GENERATION COSTS

Power Costing Models

Numerous costing models have been developed to establish the cost of producing a given amount of power from a new, typically coal-fired power plant. With a given net output level these models combine physical system parameters (i.e., mass and energy flows, see, for example, Rubin and McMichael 1978; Rubin and Nugyen 1978; Molburg 1982) with the plant operating parameters set by the utility to calculate capital cost, revenue requirements, and cost of

Table 52
Midwest Coal-by-Wire: First Year Costs

	Existing Plant and DC Line (cents/kWh)	New Plant and DC Line (cents/kWh)
Capacity Cost	0.6[a]	8.9[b]
Scrubber Cost[c]	2.9	2.9
Fuel Cost[d]	1.8	1.7
Transmission Cost[e]	5.3	5.3
TOTAL COST	10.6	18.8

Notes: [a] 1995 2600 MW plant with $245/kW installed and 1987=6 cost/kW at 3 cents.

[b] 1300 MW coal plant with $1400/kW (1987) including scrubbers. Estimated 2.5 to 3 years licensing and 4.5 to 5 years construction period. Total installed cost (1996) $3.6 billion ($2800/kW) with levelized capacity cost of $800 MM/yr.

[c] $200/kW in 1987 $.

[d] High sulfur coal.

[e] Transmission line in service January 1996 with 30-year life (2 to 4 licensing and 6 years construction). Right-of-way costs $1 million per mile. Total installed costs $2.0 billion with levelized of $425 MM/yr. Cost of 1000 MW allocated to New England with cost per kilowatt-hour of 4.9 cents (cost per kilowatt-hour based on 2000 MW, 2.5 cents).

Source: Destribats, "Power from the Midwest," 1988.

Table 53
30-Year Levelized Cost of New England Power Alternatives

	1987 Cents/kWh
New England: Combined Cycle (Gas)	6.1
New England: Combined Cycle (Oil)	7.8
New England: Coal Unit	7.6
Central Maine-Hydro Quebec	4.5
Midwest: Old Plant	5.3
Midwest: New Plant	8.6

Source: Destribats, "Power from the Midwest," 1988.

power for a series of chosen production technologies, coal characteristics, and emission regulations.

Minimizing the air-pollution control costs of alternative systems for generating electricity forms an essential part of power costing models. These costs may be dealt with either in terms of individual control options or in a range of options in some combination to meet environmental constraints.

An individual control option which has received much attention is benefication. The use of coal cleaned to a variety of levels has become an increasingly important element of the cost of conforming to emission standards at a minimum cost. The level of coal cleaning closely affects control costs and thus power costs through the varying effects coal quality has on coal transportation, handling, storage, pulverization cost, boiler fouling, post combustion cleanup, and waste disposal costs (see Phillips and Cole 1980; Roman 1980; Holt et al. 1982; Bluck and McMorris 1985).

J. C. Molburg and E. S. Rubin (1983) have developed a comprehensive series of economic models to assess the costs of conforming with emission regulations for a range of pollution control options and drawing from a wide range of technical surveys. Options considered are coal cleaning (see also Bechtel Power Corporation 1977; Molburg 1982), dry fly ash collectors (see also Bubenick 1978; Farber 1978, 1981; U.S. Environmental Protection Agency (EPA) 1977), wet limestone Flue Gas Desulfurization (FGD) (see Bechtel Power Corporation 1977; Rowland et al. 1978; Torstrick et al. 1978, 1981; Barrier et al. 1978) and solid waste disposal (see Molburg 1982; Torstrick et al. 1978, 1981; Barrier et al. 1978; Anders and Torstrick 1981). The cost for each control option is established by using statistical analysis and other techniques to calculate costs when pollutant emissions, constraints, coal characteristics, capacity, and economic basis are varied. The cost of one option or combination of options is then combined with the cost of a new power plant with no pollution controls (see Buder et al. 1977; Scherer 1977; Bechtel Power Corporation 1977; Friedlander 1977) to establish the total system cost, including the economic value of plant capacity used to run the environmental control systems.

The CONSOL Coal Research Model

Overview. A number of computer models are available and are being used by industry and academia to calculate the costs of producing electric power at utilities using different assumed factor costs and usages. One such model, developed by the Consolidation Coal Company, is a comprehensive costing program designed to assess the marketability of coals of varying qualities at a range of power plant locations. This model was chosen to be used in this investigation because of its consistency with the costing procedures already described, the ease of data input and output, its ability to address issues central to this investigation and its availability and cost. Appendix

3 describes the model. The transmission costs were computed by WVU. Details of the WVU transmission model are provided in Appendix 4.

Figure 19 shows the basic structure of the CONSOL model including the two modules used to determine power costs. Operation of the model begins with combining power plant characteristics at a particular location with specified assumptions on plant size and FGD technology and coal quality characteristics. The model input then requires information on the general rate of inflation and rates of inflation for capital expenditures, plant operating labor, barge and rail transportation, and coal. The escalation assumptions used in the model affect the levelized bus-bar power costs and also affect the relative costs of power produced by using different coals. Capital and operating costs are then calculated in the economics module. Assumed values are used for the financial variables, such as cost of capital, cost of debt, income tax rate, other taxes and insurance, investment tax credit, and after-tax cost of capital, and are used to compute levelized costs given a specified project schedule and plant capacity factor. The financial module then computes revenue requirements and levelized revenue requirements in current and constant dollars (mills/kilowatt-hour).

Use of the Model. Appendix 5 gives the input information used for the model runs. The cost of delivering power to each location was calculated by first incorporating the transportation rates into the model by modifying input information on coal delivered prices. This gives the cost of power at the bus bar for plants in each location. The cost of transmission for each scenario is then added to these figures using the WVU model to give the delivered cost of power at each load center.

Scenarios

Rationale. During the course of the study, over 300 scenarios were evaluated using the CONSOL Coal Research Power Costing Model and the WVU electricity transmission model. Varying the financial assumptions of the model, the characteristics, the source and price of coal, the price of the other inputs, the size of the plant, the length of time needed to construct the plant, the construction and operating characteristics of the plant, and the location of the plant, all resulted in a different price for the electricity generated. These scenarios provide a range of expected prices of power for any particular coal source and power plant and show price sensitivity to changes in selected variables. In order to maintain simplicity, the discussion is restricted to a few important scenarios.

Description. A number of coal sources exist for power plants in the vicinity of West Virginia and there are a number of power markets outside the state which might consume electricity generated in West Virginia power plants. Appendix 6 lists average coal quality characteristics for coal from northern West Virginia. The coal cost and quality figures are those suggested by the Consolidation Coal Company and appear to be reasonable for typical operations for the coal sources used. The analysis

Figure 19
Structure of the CONSOL Model

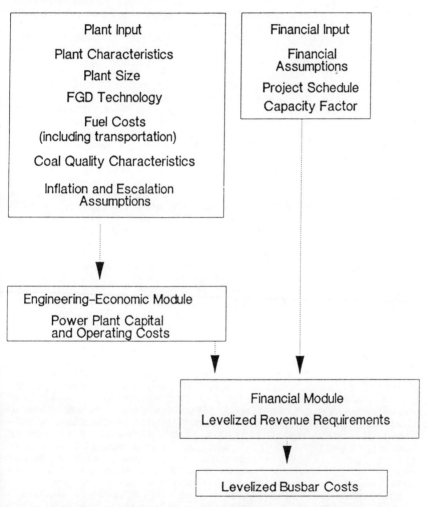

Source: CONSOL Coal Research Utility Coal Quality Power Cost Model, 1987.

is restricted to an investigation of the costs of serving the New Jersey-Philadelphia-New York City area and southern New England area with electricity. The New Jersey-Philadelphia-New York City area is served by several utilities, each with its own development plans, which complicates this analysis. To simplify this problem, we consider only the cost of delivering power into central New Jersey and assume that this power will be distributed to local demand centers. New England electricity is centrally dispatched through the NEPOOL. It is assumed that delivered power to Connecticut can be made available to serve the broad NEPOOL demands.

Power Plant Characteristics. The factors determining the costs of a plant of known location using an identified fuel are plant size, the coal burning technology used, and operating cost factors. To account for economies of scale in the construction and operation of power plants, three plant sizes were chosen for the costing analysis: 300-, 500-, and 750-megawatts (net capacity). A 300-megawatt conventional power plant is now considered small by most utilities because of the low economies of scale. However, single and multiple 300-megawatt units do present viable options in certain circumstances or when new technologies are considered. Few utilities in the United States have the demand or desire to construct a 750-megawatt fossil-fired unit for two reasons. First, a 750-megawatt plant is a large unit and only a few systems have the demand growth that would justify building a 750-megawatt plant without placing on reserve some capacity that can be economically operated. Second, the failure of a 750-megawatt plant would present major problems in systems where such a plant accounted for a large percentage of the total available capacity. Many utilities, large and small, build units in the 500-megawatt size, which appears to capture economies of scale while allowing flexibility of smaller incremental additions to capacity. A large plant can be constructed by locating several 500-megawatt plants at one site; a smaller system may build a single 500-megawatt unit as a new plant. These three sizes are reasonable to use to compare plants of different sizes.

Technology and operating cost factors also affect power generation costs. A wide range of technologies are used by power plants, particularly for the purposes of pollutant emission control. To simplify the analysis, it was assumed that all plants are newly constructed, pulverized coal-fired plants which utilize limestone scrubbers with reheat for flue gas desulfurization and electrostatic precipitators for particulate control. It is recognized that the choice of technology, particularly emission control, may well depend upon factors determined by specific site constraints and conditions. The assumption of constant technology across all sites may well mask important cost differences, but it was considered justifiable given the purposes of this investigation. Appendix 7 summarizes power plant technology characteristics for a 300-, 500-, and a 750-megawatt plant. Plant capacity utilization rates of 65 percent and 80 percent were used. FGD waste was disposed by forced oxidization at a specified cost of $10 per ton. Plant cost factors used in the model were those provided by Steans Catalytic Corporation (1984) and were computed by multiplying together costs, labor rates, and productivity.

Financial and Inflation Assumptions. The cost of project financing and the prevailing rate of inflation have a great impact on the final cost of power. Although a variety of finance and inflation scenarios were analyzed, the investigation limited itself to assuming a 6-percent rate of inflation, a 6-percent cost of debt, and a 6-percent cost of capital. It was felt that a 6-percent inflation rate was reasonable, and in any case, inflation does not affect relative economies when the same rate is assumed for all plant sites. A cost of capital of 6-percent was also considered reasonable given the prevailing levels of rate of return in the utility industry.

Coal Transportation Assumptions. The comparison between mine-mouth plants in West Virginia to plants in New Jersey includes the price of coal at the mine (f.o.b.) plus the coal transport price to deliver the coal from the mine to the plant. It is very difficult to obtain coal transportation rates because these are negotiated between the utility and the rail company. We were able to obtain transportation contract information that was collected by the West Virginia Public Service Commission for selected shipments from West Virginia to destinations within and outside of the state. This source was used to determine the transportation rates for the West Virginia mine-mouth plants and for the shipments from West Virginia to New Jersey. Coal transportation rates to New England were approximated by information provided to us by New England utilities on their current rail shipments.

In the case of the West Virginia mine-mouth plant, coal is delivered by rail at a cost of $5.77 per ton, which is an actual contract rate for a West Virginia power plant. Although other power plants in the state have lower coal delivery rates (particularly if they are supplied by conveyors or barge), the $5.77 rate is prudent to use for a new hypothetical power plant. A rate of $18.86 per ton is used for shipping coal from West Virginia to New Jersey, an actual rate derived from the West Virginia Public Service Commission; a rate of $20.00 per ton is used for transporting coal from West Virginia to New England, an actual rate taken from a New England utility.

Transmission Line Assumptions. Transmission costs were calculated for the transfer of power from West Virginia to New Jersey and Connecticut over a 500-kilovolt AC line at three power levels: 300-, 500-, and 750-megawatts. Upper and lower bounds of transmission costs were calculated to reflect the variation in capital cost over the length of the line due to variations in the cost of acquiring right-of-way.

Results of the Comparative Cost Analysis. Tables 54 and 55 show under various assumptions the bus-bar cost comparisons for a new plant located in West Virginia using coal from the northern West Virginia field and transmitting power to New Jersey and Connecticut, versus a plant located in the New Jersey and Connecticut load centers importing coal from the northern West Virginia field. Each table compares bus-bar costs from plants of three sizes (300-,

Table 54
Comparative Cost Results for a New Plant in New Jersey vs. Power from West Virginia
(mills/kWh current 1985 dollars)

Plant Location	Plant Size (MW)	Capacity Factor (%)	New Plant Capital Cost	Energy Cost	Scrubber Cost	Bus Bar Cost[a]	Transmission Cost Scenarios[b] Low	Transmission Cost Scenarios[b] High	Delivered Cost Scenarios[b] Low	Delivered Cost Scenarios[b] High
New Jersey	300	65	26.23	60.90	18.48	119.59	---	---	119.59	
		80	21.32	60.90	15.93	109.67	---	---	109.67	
West Virginia	300	65	23.66	42.73	18.15	97.88	15.67	40.96	113.55	138.84
		80	19.22	42.73	15.65	88.62	15.57	40.86	104.19	129.48
New Jersey	500	65	22.17	60.90	15.99	110.58	---	---	110.58	
		80	18.02	60.90	13.90	102.35	---	---	102.35	
West Virginia	500	65	19.94	42.73	15.71	89.39	10.61	25.84	100.00	115.23
		80	16.2	42.73	13.67	81.72	10.47	25.71	92.19	107.43
New Jersey	750	65	19.59	60.89	15.71	106.02	---	---	106.02	
		80	15.92	60.89	13.68	98.64	---	---	98.64	
West Virginia	750	65	17.57	42.73	15.42	85.12	8.55	18.76	93.67	103.88
		80	14.28	42.73	13.44	78.26	8.37	18.58	86.63	96.84

  N.WV-NJ N.WV-WV
Coal Transportation Cost: $18.85/ton $5.77/ton N.WV Coal Price: $25/ton

Inflation: 6%   Cost of Debt: 6%   Cost of Capital: 6%

	WV	NJ	Construction Schedules:	WV	NJ
Power Plant Capital Costs: 300MW	$963/kW	$1050/kW			
500MW	$799/kW	$ 871/kW	Land Purchase	1/1987	1/1987
750MW	$711/kW	$ 775/kW	Startup	6/1994	6/1994

[a]Total includes plant consumables and operation and maintenance.

Transmission right-of-way costs (500-kV line): Low: $506,000/mile; High: $1,520,320/mile.

Source: CONSOL Coal Research Power Costing/Coal Quality Model.
  [b]Author' calculations.

500-, and 750-megawatts) and uses a high and low transmission line cost for a 500-kilovolt line to allow for variations in land acquisition costs between West Virginia and the east. Also, 65- and 85-percent capacity factors are evaluated. The delivered cost of West Virginia power at the load center is compared with locally generated power, and the advantage or disadvantage to West Virginia plants is indicated. Coal prices, coal transportation rates, financial and economic assumptions, details of power plant capital cost assumptions, and transmission line costs are listed at the foot of each table. These tables are in

Table 55
Comparative Cost Results for a New Plant in Connecticut vs. Power from West Virginia
(mills/kWh current 1985 dollars)

Plant Location	Plant Size (MW)	Capacity Factor (%)	New Plant Capital Cost	Energy Cost	Scrubber Cost	Bus Bar Cost[a]	Transmission Cost Scenarios[b] Low	Transmission Cost Scenarios[b] High	Delivered Cost Scenarios[b] Low	Delivered Cost Scenarios[b] High
Connecticut	300	65	24.65	62.50	18.12	118.79	---	---	118.79	
		80	20.03	62.50	15.64	109.38	---	---	109.38	
West Virginia	300	65	23.66	42.73	18.15	97.88	22.80	60.85	120.68	158.73
		80	19.22	42.73	15.65	88.62	22.65	60.71	111.27	149.33
Connecticut	500	65	21.00	62.49	15.69	110.40	---	---	110.40	
		80	17.06	62.49	13.67	102.57	---	---	102.57	
West Virginia	500	65	19.94	42.73	15.71	89.39	15.50	38.47	104.89	127.86
		80	16.2	42.73	13.67	81.72	15.29	38.27	97.01	119.99
Connecticut	750	65	18.68	62.49	15.41	106.19	---	---	106.19	
		80	15.18	62.49	13.44	99.14	---	---	99.14	
West Virginia	750	65	17.57	42.73	15.42	85.12	12.57	28.00	97.69	113.12
		80	14.28	42.73	13.44	78.26	12.30	27.73	90.56	105.99

Coal Transportation Cost: N.WV-CT $20.00/ton   N.WV-WV $5.77/ton   N.WV Coal Price: $25/ton

Inflation: 6%   Cost of Debt: 6%   Cost of Capital: 6%

Power Plant Capital Costs:
	WV	CT
300MW	$963/kW	$956/kW
500MW	$799/kW	$794/kW
750MW	$711/kW	$708/kW

Construction Schedules:
	WV	CT
Land Purchase	1/1987	1/1987
Startup	6/1994	6/1994

[a]Total includes plant consumables and operation and maintenance.

Transmission right-of-way costs (500-kV line): Low: $506,000/mile; High: $1,520,320/mile.

Source: CONSOL Coal Research Power Costing/Coal Quality Model.
[b]Author' calculations.

nominal dollars. Appendix 8 summarizes the results for all of the tables derived in this cost analysis using constant dollars.

Table 54 compares the bus-bar costs for new coal-fired power plants in New Jersey to the cost of power delivered from new coal-fired power plants in West Virginia, assuming that both plants have identical construction schedules. Plants in both locations are begun in January 1987 and completed 7.5 years later in June 1994. The table considers 300-, 500-, and 750-megawatt plants operating at 65- and 80-percent capacity factors. The f.o.b. coal price is $25 per ton, and the transportation cost of coal from West Virginia to New Jersey is $18.85 per ton. The cost of transmission is calculated using low and high costs of $506,000 per mile and $1,520,320 per mile. The U.S. Department of Energy (DOE) evaluated the total costs of a number of recent and proposed transmission lines in their study of Canadian power imports (U.S. Department of Energy 1987). Interpolating from available data from Northern States Power and adding a premium for Quebec's topography, the DOE estimates the total cost of a Hydro Quebec 315-kilovolt (AC) line at $419,200 per mile. The high estimate used in the WVU study is close to that incurred by the 345-kilovolt (AC) New York Power Pool line that cost $1.5 million per mile.

A number of observations can be made concerning the competitiveness of coal-by-wire from Table 54. First, the total bus-bar costs are much lower when the plant operates at an 80-percent capacity factor rather than at 65-percent. At the higher rate, the capital costs of the plant and scrubber are reduced by being spread over greater electricity sales. Second, coal-by-wire is clearly cheaper when a 750-megawatt plant is considered, is competitive when a 500-megawatt plant is evaluated, and is probably more expensive for a 300-megawatt plant. For a 750-megawatt plant at 65-percent capacity, the delivered cost of power from a plant located in New Jersey is 106.02 mills/kilowatt-hour compared to a range of from 93.67 to 103.88 mills/kilowatt-hour (depending upon right-of-way costs) for a plant located in West Virginia with the power transmitted to New Jersey. The savings for a coal-by-wire scheme could be as low as 2.14 mills/kilowatt-hour to as high as 12.35 mills/kilowatt-hour. For the same size plant at an 80-percent capacity factor, the New Jersey plant's cost is 98.64 mills/kilowatt-hour to the West Virginia plant's cost of from 86.63 to 96.84 mills/kilowatt-hour. At this level, the savings could range from 1.8 mills/kilowatt-hour to as much as 12.01 mills/kilowatt-hour. It is doubtful that the cost of the right-of-way would be in the $1.5-million-per-mile area.

In the case of the 300-megawatt plant at a capacity factor of 65 percent, the West Virginia plant cost will range from 113.55 to 138.84 mills/kilowatt-hour compared to 119.59 mills/kilowatt-hour for a plant located in New Jersey. With low transmission right-of-way costs, coal-by-wire is cheaper by 6.04 mills/kilowatt-hour, but at the higher right-of-way costs, coal-by-wire is more expensive by 19.25 mills/kilowatt-hour. In both cases, the cost of electricity produced is very expensive compared to other alternatives: it is doubtful that coal-by-wire would be attractive for this size plant. The transmission costs are very high at this level, and probably it is not realistic for 300-megawatts to be transmitted over a 500-kilovolt line for this distance. For the 500-megawatt plant, the cost of the New Jersey plants (for both the 65- and 80-percent capacity factors) lies toward the center of the range of the costs for the West Virginia plants.

Table 55 summarizes the comparison of new power plants in West Virginia and Connecticut to serve the New England market with electricity. As in the New Jersey case, a high and low transmission cost is considered as well as the three different plant sizes and the 65- and 80-percent capacity factors. The outlook for coal-by-wire to New England does not appear to be as favorable as that to New Jersey. This results only from the additional transmission costs, since the additional problems of obtaining the rights-of-way through New York to reach Connecticut are not considered. For the 300-megawatt plant, the cost of delivering power from West Virginia appears to be prohibitively high, ranging from 120.68 mills/kilowatt-hour to 158.73 mills/kilowatt-hour in the case of a plant operating at a 65-percent capacity factor and from 111.27 mills/kilowatt-hour to 149.33 mills/kilowatt-hour for a plant operating at an 80-percent capacity factor. The bus-bar costs for a plant in Connecticut using $25-per-ton West Virginia coal would be 118.79 mills/kilowatt-hour for the 65-percent capacity factor plant and 109.38 mills/kilowatt-hour for the plant operating at an 80-percent capacity factor.

The 750-megawatt plant, using the West Virginia location could possibly provide savings under certain circumstances, but overall it appears to be less expensive to generate power in New England to serve that market. For a plant operating at a 65-percent capacity factor, the bus-bar cost for a New England plant would be 106.19 mills/kilowatt-hour compared to a range of from 97.69 to 113.12 mills/kilowatt-hour for a plant located in West Virginia. The midpoint of the West Virginia bus-bar costs is almost identical to the cost of generating power at the Connecticut location. In the case of the 500-megawatt plant, the West Virginia location appears to be competitive to the Connecticut plant only if the lower right-of-way costs are used. The cost of waste disposal in Connecticut will also likely be much higher than in West Virginia, if waste disposal sites can be found. This would certainly drive up the cost of generating power in Connecticut.

In summary, coal-by-wire appears to be an option worthy of further investigation for New Jersey, particularly at larger plant sizes, but it is doubtful that new plants in West Virginia could compete with new plants located in New England to serve the New England market. These costs, however, assume the construction of new AC transmission facilities. If new transmission facilities are not needed, then coal-by-wire is more attractive than the local option in all cases. Also, the cost of DC transmission may be less expensive over the distances considered than is AC transmission so that coal-by-wire schemes using DC high-voltage transmission lines could be less expensive than those portrayed here. It is possible that 300-megawatt of capacity could be added in West Virginia and the power could be transmitted to New Jersey or New England without any additional transmission line construction. Prudent planning, however, would involve the provision of additional transmission facilities to guarantee the availability of the lines for any coal-by-wire scheme.

It may be possible to approve and construct new power plants in West Virginia more quickly than equivalent plants in New Jersey or New England for the following reasons. First, new plants in West Virginia could be built by the West Virginia Public Energy Authority without needing state public service commission licensing. Once completed, the facility could be operated by an electric utility. Second, our public opinion polls indicate that West Virginia residents are favorably disposed toward coal-fired power plants in West Virginia and generally would like to have new plants located in the state. Given the level of support found, it is unlikely that there would be widespread citizen opposition to new power plant projects in the state. Finally, recently constructed power plants were licenced and constructed without incident, were completed very close to schedule, and were strongly supported by the local communities. Given these reasons, we now evaluate the comparative costs of a coal-by-wire scheme in which new power plants are completed in West Virginia more quickly than equivalent plants in New Jersey. Compressing the time between the initial construction of the plant to the point that it comes on-line would shorten the capital debt period, thus lowering the costs which are passed into the rate base.

This scenario was tested assuming that a West Virginia plant could be constructed two years sooner than a similar plant in New Jersey (5.5 years versus 7.5 years). Table 56, which gives power costs using an accelerated construction schedule for plants in West Virginia, shows that West Virginia power is favorably priced at the New Jersey load center for both 500-megawatt and 750-megawatt plants at both the upper and lower bounds of transmission costs. For the lower bound, all three power levels have an advantage, which increases from 17.3 mills/kilovolt-hour for the 65-percent capacity factor, 300-megawatt plant to 22.2 mills/kilowatt-hour for the 65-percent capacity factor, 750-megawatt plant. For the high-cost transmission scenario, the 300-megawatt (65-percent capacity factor) West Virginia plant is 8 mils/kilowatt-hour more expensive than the New Jersey facility. In the case of the 750-megawatt plant (65-percent capacity factor), the West Virginia plant is still 12 mills/kilowatt-hour less expensive than the New Jersey plant. Table 57 considers an accelerated construction schedule in West Virginia compared to a conventional schedule for a new plant in Connecticut. For the low transmission cost rate, it is less expensive to build a new plant and transmission line (500 kilovolts) to Connecticut than to build a load center plant. This is true for the 300-, 500-, and 750-megawatt plant sizes. If the higher transmission costs are used, which should be done in the case of building new lines to New England, only the 750-megawatt plant (at both 65- and 80-percent capacity factors) is the marginally cheaper alternative if constructed in West Virginia.

Electric power plant construction schedules have had a history of becoming protracted in the eastern United States because of unexpected environmental objections. We have seen that power plants sited in West Virginia were ordered and brought on-line rather quickly. It is possible, therefore, that not only could a West Virginia plant be constructed two years earlier because of reduced state requirements, but also a plant in New Jersey or Connecticut could be delayed for a number of years given the more

Table 56
Comparative Cost Results for a New Plant in New Jersey vs. Power from West Virginia:
  Two-Year Accelerated Construction Schedule for West Virginia Plant
  (mills/kWh current 1985 dollars)

Plant Location	Plant Size (MW)	Capacity Factor (%)	New Plant Capital Cost	Energy Cost	Scrubber Cost	Bus Bar Cost[a]	Transmission Cost Scenarios[b]		Delivered Cost Scenarios[b]	
							Low	High	Low	High
New Jersey	300	65	26.23	60.90	18.48	119.59	---	---	119.59	
		80	21.32	60.90	15.93	109.67	---	---	109.67	
West Virginia	300	65	20.76	38.03	16.07	86.73	15.55	40.89	102.28	127.62
		80	16.86	38.03	13.87	78.57	15.46	40.76	94.03	119.33
New Jersey	500	65	22.17	60.90	15.99	110.58	---	---	110.58	
		80	18.02	60.90	13.90	102.35	---	---	102.35	
West Virginia	500	65	17.50	38.03	13.91	79.25	10.43	25.66	89.68	104.91
		80	14.22	38.03	12.11	72.48	10.31	25.55	82.79	98.03
New Jersey	750	65	19.59	60.89	15.71	106.02	---	---	106.02	
		80	15.92	60.89	13.68	98.64	---	---	98.64	
West Virginia	750	65	15.43	38.03	13.66	75.48	8.30	18.51	83.78	93.99
		80	12.53	38.03	11.91	69.42	8.14	18.35	77.56	87.77

                            N.WV-NJ   N.WV-WV
Coal Transportation Cost: $18.85/ton $5.77/ton   N.WV Coal Price: $25/ton

Inflation: 6%   Cost of Debt: 6%   Cost of Capital: 6%

		WV	NJ	Construction Schedules:		
					WV	NJ
Power Plant Capital Costs:	300MW	$963/kW	$1050/kW			
	500MW	$799/kW	$ 871/kW	Land Purchase	1/1987	1/1987
	750MW	$711/kW	$ 775/kW	Startup	6/1994	6/1994

[a]Total includes plant consumables and operation and maintenance.

Transmission right-of-way costs (500-kV line): Low: $506,000/mile; High: $1,520,320/mile.

Source: CONSOL Coal Research Power Costing/Coal Quality Model.
         [b]Author' calculations.

complex approval procedures and possible public resistance to the plant or waste disposal facilities. Table 58 gives the bus-bar costs in a situation in which not only are new West Virginia plants able to come on-line two years ahead of the base schedule (i.e., the schedule used in Table 54), but also where New Jersey facilities take an additional two years over conventional siting schedules. At the lower bound of the transmission cost range, West Virginia power transmitted to New Jersey load centers has an advantage of between 32 and 35 mills/kilowatt-hour across the three

Table 57
Comparative Cost Results for a New Plant in Connecticut vs. Power from West Virginia:
   Two-Year Accelerated Construction Schedule for West Virginia Plant
   (mills/kWh current 1985 dollars)

Plant Location	Plant Size (MW)	Capacity Factor (%)	New Plant Capital Cost	Energy Cost	Scrubber Cost	Bus Bar Cost[a]	Transmission Cost Scenarios[b]		Delivered Cost Scenarios[b]	
							Low	High	Low	High
Connecticut	300	65	24.65	62.50	18.12	118.79	---	---	118.79	
		80	20.03	62.50	15.64	109.38	---	---	109.38	
West Virginia	300	65	20.76	38.03	16.07	86.73	22.62	60.68	109.35	147.41
		80	16.86	38.03	13.87	78.57	22.50	60.55	101.07	139.12
Connecticut	500	65	21.00	62.49	15.69	110.40	---	---	110.40	
		80	17.06	62.49	13.67	102.57	---	---	102.57	
West Virginia	500	65	17.50	38.03	13.91	79.25	15.23	38.20	94.48	117.45
		80	14.22	38.03	12.11	72.48	15.05	38.02	87.53	110.50
Connecticut	750	65	18.68	62.49	15.41	106.19	---	---	106.19	
		80	15.18	62.49	13.44	99.14	---	---	99.14	
West Virginia	750	65	15.43	38.03	13.66	75.48	12.19	27.62	87.67	103.10
		80	12.53	38.03	11.91	69.42	11.95	27.38	81.37	96.80

	N.WV-CT	N.WV-WV	
Coal Transportation Cost:	$20.00/ton	$5.77/ton	N.WV Coal Price: $25/ton

Inflation: 6%   Cost of Debt: 6%   Cost of Capital: 6%

		WV	CT	Construction Schedules:		
Power Plant Capital Costs:	300MW	$963/kW	$956/kW		WV	CT
	500MW	$799/kW	$794/kW	Land Purchase	1/1987	1/1987
	750MW	$711/kW	$708/kW	Startup	6/1994	6/1994

[a]Total includes plant consumables and operation and maintenance.

Transmission right-of-way costs (500-kV line): Low: $506,000/mile; High: $1,520,320/mile.

Source: CONSOL Coal Research Power Costing/Coal Quality Model.
        [b]Author' calculations.

plant sizes (for plants operating at 65-percent capacity). At the upper bound of the transmission costs, the advantage grows from 6.7 mills/kilowatt-hour for a 300-megawatt plant at 65-percent capacity to 25.1 mills/kilowatt-hour for power from a 750-megawatt plant. Table 59 provides the same scenario with Connecticut as in Table 58 with similarly optimistic results for West Virginia. At 300 megawatts (65-percent capacity), the West Virginia plant is 24 mills less expensive than the Connecticut plant using the lower transmission costs. With the high transmission

Table 58
Comparative Cost Results for a New Plant in New Jersey vs. Power from West Virginia:
Two-Year Accelerated Construction Schedule for West Virginia Plant, Two-Year Delayed Schedule
for New Jersey Plant
(mills/kWh current 1985 dollars)

Plant Location	Plant Size (MW)	Capacity Factor (%)	New Plant Capital Cost	Energy Cost	Scrubber Cost	Bus Bar Cost[a]	Transmission Cost Scenarios[b] Low	Transmission Cost Scenarios[b] High	Delivered Cost Scenarios[b] Low	Delivered Cost Scenarios[b] High
New Jersey	300	65	29.48	68.43	20.77	134.37	---	---	134.37	
		80	23.95	68.43	17.90	123.22	---	---	123.22	
West Virginia	300	65	20.76	38.03	16.07	86.73	15.55	40.89	102.28	127.62
		80	16.86	38.03	13.87	78.57	15.46	40.76	94.03	119.33
New Jersey	500	65	24.91	68.42	17.97	124.25	---	---	124.25	
		80	20.24	68.42	15.62	115.00	---	---	115.00	
West Virginia	500	65	17.50	38.03	13.91	79.25	10.43	25.66	89.68	104.91
		80	14.22	38.03	12.11	72.48	10.31	25.55	82.79	98.03
New Jersey	750	65	22.01	68.42	17.66	119.12	---	---	119.12	
		80	17.88	68.42	15.37	110.83	---	---	110.83	
West Virginia	750	65	15.43	38.03	13.66	75.48	8.30	18.51	83.78	93.99
		80	12.53	38.03	11.91	69.42	8.14	18.35	77.56	87.77

Coal Transportation Cost: N.WV-NJ $18.85/ton   N.WV-WV $5.77/ton   N.WV Coal Price: $25/ton

Inflation: 6%   Cost of Debt: 6%   Cost of Capital: 6%

Power Plant Capital Costs:
	WV	NJ
300MW	$963/kW	$1050/kW
500MW	$799/kW	$871/kW
750MW	$711/kW	$775/kW

Construction Schedules:
	WV	NJ
Land Purchase	1/1987	1/1987
Startup	6/1994	6/1994

[a]Total includes plant consumables and operation and maintenance.

Transmission right-of-way costs (500-kV line): Low: $506,000/mile; High: $1,520,320/mile.

Source: CONSOL Coal Research Power Costing/Coal Quality Model.
 [b]Author' calculations.

costs, it is less expensive (by 14 mills) to use the Connecticut location. At 750 megawatts (at 65-percent capacity), the West Virginia site is cheaper using both the high and low transmission cost figures.

Table 59
Comparative Cost Results for a New Plant in Connecticut vs. Power from West Virginia:
  Two-Year Accelerated Construction Schedule for West Virginia Plant, Two-Year Delayed Schedule
  for Connecticut Plant
    (mills/kWh current 1985 dollars)

Plant Location	Plant Size (MW)	Capacity Factor (%)	New Plant Capital Cost	Energy Cost	Scrubber Cost	Bus Bar Cost[a]	Transmission Cost Scenarios[b] Low	Transmission Cost Scenarios[b] High	Delivered Cost Scenarios[b] Low	Delivered Cost Scenarios[b] High
Connecticut	300	65	27.70	70.22	20.36	133.47	---	---	133.47	
		80	22.50	70.22	17.57	122.90	---	---	122.90	
West Virginia	300	65	20.76	38.03	16.07	86.73	22.62	60.85	109.35	147.41
		80	16.86	38.03	13.87	78.57	22.50	60.55	101.07	139.12
Connecticut	500	65	23.59	70.22	17.63	124.05	---	---	124.05	
		80	19.17	70.22	15.35	115.24	---	---	115.24	
West Virginia	500	65	17.50	38.03	13.91	79.25	15.23	38.20	94.48	117.45
		80	14.22	38.03	12.11	72.48	15.05	38.02	87.53	110.50
Connecticut	750	65	20.99	70.21	17.31	119.32	---	---	119.32	
		80	17.05	70.21	15.10	111.40	---	---	111.40	
West Virginia	750	65	15.43	38.03	13.66	75.48	12.19	27.62	87.67	103.10
		80	12.53	38.03	11.91	69.42	11.95	27.38	81.37	96.80

                                 N.WV-CT   N.WV-WV
Coal Transportation Cost: $20.00/ton $5.77/ton   N.WV Coal Price: $25/ton

Inflation: 6%   Cost of Debt: 6%   Cost of Capital: 6%

                                      WV   CT   Construction Schedules:
Power Plant Capital Costs: 300MW $963/kW $956/kW   WV   CT
                            500MW $799/kW $794/kW   Land Purchase   1/1987   1/1987
                            750MW $711/kW $708/kW   Startup   6/1994   6/1994

[a]Total includes plant consumables and operation and maintenance.

Transmission right-of-way costs (500-kV line): Low: $506,000/mile; High: $1,520,320/mile.

Source: CONSOL Coal Research Power Costing/Coal Quality Model.
        [b]Author' calculations.

CONCLUSIONS

The investigation of comparative cost was limited to these scenarios. The cost of power generated in a new plant in West Virginia using local coal and transmitted to eastern load centers is compared to the costs of power in a New Jersey or Connecticut load center plant which uses coal transported by rail from the West Virginia field. The cost effects of varying plant size and construction schedules were established.

Plant size was found to affect significantly the cost of power delivered to New Jersey or Connecticut from West Virginia plants. Although, at the upper bound of transmission costs, power produced in West Virginia is at a cost disadvantage compared to the cost of power produced at the load center, this turns to an advantage with larger plant sizes. This is especially true if transmission costs approach their lower bound.

Shortening the construction schedule for new plants in West Virginia has been considered a means of making power produced within the state more competitive with the load center. Our investigation confirms an increased advantage if construction schedules are accelerated by two years in West Virginia. If construction schedules are both accelerated in West Virginia by two years and lengthened in New Jersey or Connecticut by two years, the advantage to West Virginia grows considerably depending on the cost of transmission. The latter schedule reflects the likelihood of longer environmental approval procedures for additions to capacity at load center sites.

Differences in costs for a plant of any given size may vary substantially, therefore, depending on the scenarios chosen for analysis. Although these ranges are given as a guide to the dollar amounts involved, given the site-specific problems in estimating electric power costs, these ranges should not be regarded in any way as representative of actual cost differentials between plants located in West Virginia versus those located in New Jersey or Connecticut. Significant cost differentials are nevertheless available to power producers in West Virginia in competing with load-center-generated power. These cost advantages not only give the opportunity for West Virginia-based utilities to offer eastern consumers price discounts to induce the purchase of West Virginia generated power, but also they are substantial enough to provide a significant rate of return for investors considering locating generating facilities in West Virginia.

BIBLIOGRAPHY

American Electric Power Service Corporation. March 1988. Critique of Center for Clean Air Policy Report. Columbus, Ohio: American Electric Power Service Corporation.

Anders, W. L., and Torstrick, R. L. 1981. Computerized Shawnee Lime/Limestone Scrubbers Model Users Guide. Knoxville, Tenn.: Tennessee Valley Authority.

Barrier, J. W., et al. 1978. Economics of Disposal of Lime/Limestone Scrubbing Wastes: Untreated and Chemically Treated Wastes. Knoxville, Tenn.: Tennessee Valley Authority.

Bechtel Power Corporation. San Francisco Power Division. 1977. Coal Fired Power Plant Capital Cost Estimates. Palo Alto: EPRI.

Bluck, W. V., and W. L. McMorris. 1985. "Effect of Coal Preparation on Power Plant Fuel Cycle Cost Measured at the Busbar." Mining Engineering 37:1135-40.

Bubenick, D. V. 1978. "Economic Comparison of Selected Scenarios for Electrostatic Precipitators and Fabric Filters." Journal of the Air Pollution Control Association 28:279.

Buder, M., et al. 1977. Environmental Control Implications of Generating Electric Power from Coal: Coal Preparation and Cleaning Assessment Study. Argonne, Ill: Argonne National Laboratories.

Casson, W. A. 1963. "Technical and Economical Comparison Between AC and DC Transmission." Proceedings form the CIGRE Conference. Paris, France.

Dansfors, P. B. Hammarlund, and C. A. O. Peixoto. 1977. "Rapid Method for Deducing Line Costs in Relation to AC Line Costs." Paper presented at IEEE Power Meeting, New York, NY. (February).

Destribats, A. F. 1988. "Power from the Midwest." Paper presented at The New England Governors' Conference Northeast-Midwest Transmission Workshop. Nashua, N.H.

Electric Power Research Institute. 1982. Transmission Line Reference Book: 345 KV. 2nd ed. Palo Alto, CA:EPRI Press.

Farber, P. S. 1978, 1981. "Energy and Environmental Systems: Comments." Argonne, Ill.: Argonne National Laboratories.

Friedlander, G. D. 1977. "20th Steam Station Cost Survey." Electrical World. 188 (Nov. 15):43-58.

Holt, E. C., et al. 1982. "Effect of Coal Quality on Maintenance Costs at Utility Plants." Mining Congress Journal 68:48-54.

Jones, B. 1972. New Approaches to the Design of EHV Transmission Lines. Oxford:Pergamon Press.

Moallem, M. 1985. "Cost Calculation for Energy Export through EHV Transmission Lines." Morgantown, WV: Master's thesis, West Virginia University.

Molburg, J. C. 1982. "Mathematical Models for Integrated Environmental/Economic Analysis of the Coal-to-Electric Systems." Pittsburgh, PA: Ph.D. diss., Carnegie-Mellon University.

Molburg, J. C., and E. S. Rubin. 1983. "Air Pollution Control Costs for Coal-to-Electricity Systems." Journal of the Air Pollution Control Association 33:5.

Neme, C. 1987. Midwest Coal by Wire: Addressing Regional Acid Rain and Energy Problems. Washington, D.C.: Center for Clean Air Policy.

Phillips, P. J., and R. M. Cole. 1980. "Economic Penalties Attributable to Ash Content of Steam Coals." Mining Engineering 32:297-302.

Roman, C. L. 1980. "Bonification: An Alternative to 100% Flue Gas Scrubbing." Coal Mining and Processing 17:68-73.

Rowland, C. H., et al. 1978. "Predicting CO_2 Removal by Limestone/Lime Wet Scrubbing: Correlations of Shawnee Data." Paper presented at APCA Annual Meeting. Houston, TX. (June).

Rubin, E. S., and F. C. McMichael. 1978. "Cross-Media Environmental Impacts of Air Pollution Regulations for a Coal Fired Power Plant." Journal of the Air Pollution Control Association 28:1099.

Rubin, E. S., and D. C. Nugyen. 1978. "Energy Requirements of a Limestone FGD System." *Journal of the Air Pollution Control Association* 28:1207.

Scherer, R. 1977. *Estimating Electric Power System Marginal Costs*. New York: North-Holland.

Steans Catalytic Corporation. 1984. *Retrofit FGD Cost Estimating Guidelines*. Palo Alto, CA:EPRI CS-3696.

Torstrick, R. L., et al. 1978, 1981. *Emission Control Development Projects*. Knoxville, Tenn.: Tennessee Valley Authority.

U.S. Department of Energy. 1987. *Northern Lights: The Economic and Practical Potential of Imported Power from Canada* Washington, D.C.: U.S. Department of Energy.

U.S. Environmental Protection Agency. 1977. *Reports on Electrostatic Precipitator Costs for Large Coal-Fired Steam Generators*. Washington, D.C.: Environmental Protection Agency.

10

Conclusions

Frank J. Calzonetti

It should be clear to the reader that the electricity industry is an important market for the U.S. coal industry and that the fortunes of areas linked economically to coal are closely tied to the use of coal in power plants. In Appalachia this is clearly demonstrated by an investigation of the case of West Virginia. Coal purchased by out-of-state electric utilities provide important markets for coal. When markets are lost, coal mine output, coal mine employment, and state revenues fall. When coal is burned at West Virginia power plants, the state gains additional revenues and jobs. Revenues are gained at the state and local level.

It should also be recognized that electricity should be viewed as a product as well as a service. The manner in which electric utilities are regulated has focused attention on the service role of electricity. Utilities must be able to provide reliable and economical service to their customers. The fact that electricity can also be an important product must not be overlooked, nor should this role be viewed disparagingly. Areas such as Appalachia that are able to generate inexpensive power should do so even if that power is exported beyond the region. Appalachia gains in many ways from this sale of power, and we see no way in which customers in the regions suffer because of this.

We have also seen that under many assumptions, importing power from plants located in West Virginia can be less expensive than transporting coal to plants located in New Jersey, and in some cases can be less expensive than generating power at new plants located in Connecticut. This even assumes that new transmission facilities are constructed to deliver this power. Furthermore, our surveys show that most people in West Virginia are favorably disposed toward power plants. The technology is familiar to them and it is largely considered benign. Also, most West Virginians living near power plants voice little concern about their operations and recognize the substantial positive impact that they bring to the local and state economy. Most of the people surveyed would like to see additional coal-fired power plants located in West Virginia. It is doubtful that large coal-fired power plants could be easily sited in the East. In addition to public opposition, it would be extremely difficult to locate approved landfills for fly ash and scrubber sludge.

Our results are very positive about the electricity industry in the region. We are not sure whether there will be any major expansion of the region's electricity-exporting capability. Although coal-fired power plants in Appalachia are among the most efficient in the world, although there is a strong desire by state officials and coal interests to expand the West Virginia electricity export capability, and although local citizens view the idea favorably, the ability of the region to increase its electricity-exporting capability is limited by institutional and geographical factors as well as major uncertainties facing the electricity industry. There are encouraging signs that favor increased coal-by-wire development: the Moore-Sununu initiative, the National Governors' Association Electricity Transmission Task Force, and the proposed Pennsylvania Electric Power Transmission Task Force. Also, the robust electricity demand in the East, coupled with problems in bringing nuclear capacity on-line, is forcing utilities to consider all options.

The organization of the electricity industry does not encourage the siting of power plants outside their service districts. Legally, most electric utilities would be allowed to build or partly own a power plant in another region. Many utilities are hesitant to do this. Although such a facility could probably be added to the rate base, it might be difficult for a utility to gain eminent domain rights for the power plant, substations, and, most important, transmission lines. Thus, even if an electric utility were interested in building new capacity, the manner in which electric utilities are organized and regulated works against them in building capacity outside of their service districts. The power company would need to seek rate hikes from a public utility commission in the state that it serves on a facility located perhaps in a state where it has no customers. In addition, most power companies are interested in promoting the economic growth in their service districts. Most power company officials realize the economic benefits that new energy facilities bring to an area and would prefer to see that development occur in their own region.

Control of energy facilities is another factor. A power plant located in another state would be subject to the control of that state's governor and legislature. Locating the plant in the state might be economical today, but the state could pass legislation (e.g., electricity severance tax) that would alter the economics greatly. Many electric utilities work closely with officials in the state capitol, and they would not have this influence over facilities located elsewhere.

The uncertainty surrounding the electricity industry makes it difficult to plan an electricity-exporting strategy. The greatest uncertainty is in the demand for electricity. Utilities were severely hurt by overbuilding in response to an expected robust electricity demand growth that did not materialize. Utilities are prepared to wait until the electricity demand situation clarifies before committing themselves to a construction strategy. The problem with this is that a few years of high growth levels can drop reserve margins considerably, as was shown in Chapter 3. We are seeing this in the northeastern United States. We do believe that it will be necessary for the Mid-Atlantic Area Reliability Council (MAAC) and the New England Power Pool (NEPOOL) regions to expand their capacity in the next decade in order to maintain industry standards of reliability. Energy from the Appalachians should be considered as one alternative to increase the power supplies to these regions. Utilities in the MAAC region, in particular, should engage in contracts with utilities or other entities in the Appalachians or Midwest for a portion of their future supply requirements.

Our analysis shows that building new power plants in West Virginia will probably be less expensive than building power plants in New Jersey, but the power from these facilities will not be as inexpensive as the power now being generated at existing West Virginia power plants. If the demand for power does not materialize and if contracts are not written, new power plants probably will not be competitive with the existing capacity in the region. In addition, the state gains by expanding the utilization of existing facilities in many ways.

There is also uncertainty concerning which type of fuel to use to meet growing needs in the East. We believe that coal is still the best choice for new base load facilities, but many choices are available in coal technologies. Currently, the pulverized coal-fired power plant, with a scrubber, is the primary choice for sizes of 500 megawatts and larger. At a smaller scale, there are other attractive alternatives, particularly fluidized-bed combustion and integrated combined-cycle gasification. These technologies are being given serious consideration by utilities.

Acid rain legislation is a serious question for the Appalachian region, particularly since so much of the capacity lacks scrubbers. A requirement to reduce sulfur oxide emissions could make these power plants much less competitive and could reduce the region's electricity export capability. We expect that the region will be required to address this problem by federal action. Requiring existing coal-fired facilities to install scrubbers will, of course, reduce the present-day competitiveness of electricity from the Appalachians and Midwest. This would probably reduce the amount of economy energy being traded from The East Central Area Reliability Council (ECAR) to the MAAC, reduce the revenues to electricity-exporting regions, diminish the positive indirect impacts of operating plants at a higher rate. However, if the need is to purchase capacity, then the impact of acid rain legislation will be less severe if new coal-fired facilities are being compared to other new capacity alternatives.

Canadian power imports also pose a threat to the region's ability to export power to the Northeast. The level of imports will rise but will not eliminate the need for the Northeast to purchase power from the Appalachians.

Even though hydroelectric power from Canada is less expensive than power from the coalfields, utilities are seeking a diverse mix of electricity supplies for reliability and economic risk reasons. There is concern that Hydro Quebec has too many outages. Also, the power is not under the control of a U.S. utility, so that price or availability could possibly change in the future, despite trade agreements. U.S. utilities could have more control over power supplies originating in another state through the federal government.

The siting of new transmission lines to bolster power transfer across the Appalachians appears to be needed. Unless new mechanisms, such as multistate compacts, power marketing areas, or state-initiated programs, are established, these lines will be difficult to build. Although there is no concern over the reliability of the interconnected system at this time, it should be recognized that the system now in place was not designed for large-scale continuous power trade at current levels. State governments should work together to expand new transmission facilities to make the best use of existing price and reserve margin disparities and to prepare for future electricity trade.

It appears that, if coal-by-wire schemes are evaluated as individual projects on their economic merit alone, it is doubtful whether many utilities would select this option for meeting their future requirements. The economics are not that good. However, the need to construct new transmission lines could be based on regionwide economic, reliability, and diversity criteria and not evaluated as a particular energy project. New transmission facilities could have far-reaching benefits to the northeastern United States well into the middle of the next century. The evaluation of new transmission facilities on these bases would greatly enhance the attractiveness of coal-by-wire.

The electricity industry was developed at a local scale at first, then was expanded as service areas were developed at the regional scale. The power blackout of 1965 led power planners to expand electricity reliability planning to the multistate scale. Multistate power planning and cooperation can lead to savings to customers in electricity-importing regions and revenues and jobs to people in electricity-exporting regions. Also, power plants could be constructed in areas where people would prefer to have them.

Our conclusions, however, are tempered by the knowledge that insufficient data are available on the ability of the interconnected system to accommodate increased transfers of power if selected new lines are constructed. We do know that it is very expensive to build new transmission lines and that it may be difficult to win support for the construction of these lines. Without additional load flow studies of the system, we cannot know how investments in new facilities can alter the exchange of power under a coal-by-wire strategy. We believe that the electric utility industry should join with federal and state officials to begin planning for regionwide transmission systems designed to accommodate long-term capacity and economy transactions.

Appendixes

Appendix 1:
Comparison of Control Counties to Impact Counties

Table A-1.1
Calibration Period Test of the Mason County Control Group

Sector	Actual Change 1970-1975 ($1,000s)	Impact
Farming & Agricultural Services	-532	-824
Mining	957	-1003
Construction	1016	-595
Manufacturing	5799	1681
Transportation, Communication, and Public Utilities	2759	-1099
Wholesale Trade	2115	345
Retail Trade	1255	183
Finance, Insurance, and Real Estate	500	42
Services	2636	447
Federal	614	-10
State and Local Government	2348	-796
Residential Adjustment	5494	1729
Transfer Payments	4117	-524
Dividends, Interest, Rents	10851	54
TOTAL	39929	-370

Table A-1.2
Calibration Period Test of the Pleasants County Control Group

Sector	Actual Change 1967-1973 ($1,000s)	Impact
Farming & Agricultural Services	-108	-418
Mining	-5	-18
Construction	3412	3058
Manufacturing	4423	-441
Transportation, Communication, and Public Utilities	282	-1069
Wholesale Trade	78	69
Retail Trade	369	-203
Finance, Insurance, and Real Estate	119	-56
Services	546	-61
Federal	153	26
State and Local Government	2196	-1211
Residential Adjustment	-2444	211
Transfer Payments	1200	-95
Dividends, Interest, Rents	2316	42
TOTAL	12537	-166

Table A-1.3
Calibration Period Test of the Grant County Control Group

Sector	Actual Change 1959-1962 ($1,000s)	Impact
Farming & Agricultural Services	-996	-489
Mining	-60	-69
Construction	-5	-8
Manufacturing	33	-43
Transportation, Communication, and Public Utilities	18	-5
Wholesale Trade	205	108
Retail Trade	-29	-16
Finance, Insurance, and Real Estate	29	6
Services	30	-11
Federal	94	18
State and Local Government	212	67
Residential Adjustment	-216	-150
Transfer Payments	217	97
Dividends, Interest, Rents	305	29
TOTAL	-109	-466

Table A-1.4
Calibration Period Test of the Putnam County Control Group

Sector	Actual Change 1962-1968 ($1,000s)	Impact
Farming & Agricultural Services	-182	-206
Mining	-181	-144
Construction	2750	848
Manufacturing	6050	284
Transportation, Communication, and Public Utilities	998	-220
Wholesale Trade	135	37
Retail Trade	708	-75
Finance, Insurance, and Real Estate	215	123
Services	1266	837
Federal	91	20
State and Local Government	1312	-131
Residential Adjustment	10119	-676
Transfer Payments	1651	-851
Dividends, Interest, Rents	2425	27
TOTAL	27357	-1

Appendix 2:
Data Used for AEP Coal-by-Wire Estimates

Energy cost calculations are based on actual cost of fuel and variable O & M at Rockport Plant for 1987.
Demand cost calculations are based on cost of Rockport #2 ($907/kW and fixed O & M @ $9.25/kW (1987 $) @ 17.83%.
Scrubber capital cost based on installed cost of scrubbers at Gavin plant of $250/kW @ carrying charge rate of 17.83%.
Scrubber O & M based on total O & M cost of scrubbers @ $95 million per year.
Scrubber replacement capital (1995) based on replacement of 6.5% (169 MW) of capacity "lost" due to addition of scrubbers-assumed at $907/kW (Rockport #2) @ carrying charge rate of 17.83%.
Scrubber replacement O & M (1995) based on fixed O & M cost of $9.25/kW Rockport (1987 $) for replacement of 6.5% of "lost" capacity.
Scrubber replacement capital (2000) based on permanent replacement of 6.5% (169 MW) of "lost" capacity at installed cost of $1570/kW (assumed to be PFBC capacity (1987 $) @ carrying charge rate of 17.83%.
Scrubber O & M (2000) based on PFBC capacity fixed O & M of $21.90 per kW/yr and replacement of 6% of "lost" capacity.
New plant capacity capital cost (2000) based on new plant capital cost of $1570/kW (2,600 MW PFBC units, 1987 $) @ carrying charge rate of 17.83%.
New plant capacity O & M (2000) based on PFBC fixed O & M cost of $21.90 kW/yr.
Power line costs as given by CCAP (100% load factor @ 17% carrying charge).
Delivered cost to New England assumes 5% transmission losses on the HVDC line.

Appendix 3:
Power Costing Analysis

A consistent method of cost calculation is required to generate valid comparisons between competitive power generation options. The following section describes the technique used to evaluate power costs in this analysis.

Levelized busbar power costs are used to measure economic effectiveness. They represent the life-cycle costs (including both capital and operating contributions) of an electrical generation project as a single value. These costs are calculated per unit of generated power and include the impacts of coal quality, plant size, and plant location.

Levelized power costs are calculated in three steps. First, the site-specific cost of plant construction (including financing) is calculated yielding the total capital cost or undepreciated rate base. Second, the revenue required to cover operating expenses and capital charge (income to recover capital cost, obtain a suitable return on investment, and fulfill tax obligations) is determined for each year of plant operation. Finally, the year-by-year costs are discounted to the initial year of operation and divided by the quantity of power generated to yield the levelized busbar costs.

Total Capital Costs. Capital investments are divided into four categories: construction, replacement equipment, land, and fuel inventory. The accounting method for each category is slightly different with regard to book value, purchase date, etc. The undepreciated rate base for a given asset is the book value of that asset in that year.

The construction cost includes capital cost for direct material and labor plus the indirect and distributable costs. An allowance for funds used during construction (AFUDC) is allocated based on an estimated plant investment centroid of two years prior to plant start-up and using an interest rate equal to the after tax cost of capital. The book value decreases linearly over the life of the plant.

A zero salvage value is assumed. Prior to 1986, an investment tax credit was allowable for construction costs. If applicable, the investment tax credit is credited during the first year of operation.

Capital costs for replacement equipment are treated similarly to plant construction costs, except that there is no allowance for funds used during construction and the book value decreases over the period of use (three or ten years after the date of purchase). The value for replacement equipment is escalated to the year in which it is installed. Prior to 1986, investment tax credit was permitted for ten year equipment but not for three year equipment.

Land includes the cost for the total area needed for the power plant including storage and disposal areas. It is not depreciated but carried at acquisition cost plus the interest accumulated prior to commercial operation. Land is assumed to have a salvage value of 100% and no investment tax credit is allowed.

Fuel inventory, which includes both acquisition and transportation costs, is treated as working capital. The fuel inventory is financed each year with the value increasing at a rate equal to the escalation risk of the delivered coal price. There is no depreciation or investment tax credit for this category of the capital cost.

Operating Cost. Operating costs are divided into five categories: fuel (including transportation), consumables, operation and maintenance labor, operation and maintenance materials, and administration and supervision. Fuel is by far the most significant operating cost, consisting of 81% of operating cost for a plant with flue gas desulfurization (Consol R&D Power Costing Model). The cost of consumables varies according to the amount of scrubbing employed. In addition, economies of scale may exist in some elements of the operating costs.

Total Revenue Requirements. The revenue requirements needed to cover all capital charges and operating expenses are calculated for each year of the project. Capital charges represent income needed to pay off the capital investment, obtain a suitable return on equity and fulfill tax obligations.

The operating expenses are specified in constant dollars. Their contribution to the revenue requirements in any given year is adjusted for escalation and, where appropriate, the unit capacity factor.

The capital charges for an asset in any year are defined as follows:

$$CC = BD + ROE + TX + INS \tag{1}$$

where:

CC = Capital Charges
BD = Book Depreciation
ROE = Return on Equity
TX = Income Tax Burden
INS = Other Taxes and Insurance

Book depreciation, BD, is income which is applied to the reduction of the book value of a depreciable asset. Book depreciation for non-depreciable assets is zero. Return on equity, ROE, is calculated by multiplying the rate of return by the book value of the asset (undepreciated rate base). The rate of return is equal to the after tax cost of capital (COC). Incomes for book depreciation and return on equity incomes are taxable and therefore require additional revenues. This revenue is the income tax burden, TX. Some relief from the income tax burden is obtained from accelerated depreciation, called tax depreciation, TD. Tax depreciation is based on the installed cost and is calculated using a schedule defined by the tax code over the tax life of the asset. Prior to 1986, a fifteen year tax life was used for power plant construction costs. Replacement equipment assets with book lives of less than fifteen years use tax lives equal to their book lives.

The tax burden can be calculated as follows:

$$TX = (TR/(1 - TR)) * (ROE + BD - TD) \quad (2)$$

where:

TX = Income Tax Burden
TR = Tax Rate
TD = Tax Depreciation

The investment tax credit where appropriate is calculated as follows:

$$ITC = (ITR/(1 - TR)) * BV \quad (3)$$

where:

ITC = Investment Tax Credit
ITR = Investment Tax Credit Rate
BV = Book Value

The investment tax credit is applicable only during the first year of asset life. The total capital charge is adjusted by subtracting the investment tax credit.

Other taxes and insurance are calculated as a fixed percentage of the maximum book value of the asset. Immediate flow through of tax benefits is assumed for both the income tax burden and investment tax credit. Other taxes and insurance expenses are tax deductible and therefore, do not affect the tax burden.

<u>Levelized Revenue Requirements</u>. Levelizing the year-by-year cost streams is accomplished by calculating their present value. The cost of capital to the utility is used as the discount factor. Revenues are discounted to the year of initial operation using the present worth factor (PWF) defined as follows:

$$PWF_y = 1/(1+COC/100)^y \quad (4)$$

where:

PWF_y = The present worth factor for year y
COC = The after tax cost of capital
y = Year of operation

The levelized revenue requirements are expressed as levelized annual costs in current dollars. The equation for calculating these values is as follows:

$$CURLEV = TPV / \sum_{y=1}^{life} PWF_y \qquad (5)$$

where:

CURLEV = Current dollar levelized revenue requirements, dollars
TPV = Total present value

A more useful measure of levelized revenue requirement is the current dollar generation cost which represents the busbar cost of power per unit of plant output based on a projected plant capacity factor. This is calculated using the following equation:

$$CURGEN = TPV \; KWHR * \sum_{y=1}^{life} PWF_y * CF_y \qquad (6)$$

where:

CURGEN = Constant dollar levelized generation cost, mill/kWh
KWHR = Annual power generation at 100% capacity
CF_y = Capacity factor in year y

Appendix 4:
WVU Electricity Transmission Model

The total cost of an electric transmission line consists of initial costs (capital costs) and cost of power losses and maintenance (operating costs) occurring throughout its life. For accounting and pricing purposes, the overall cost of the transmission line is calculated by capitalizing the operating cost to its present worth and adding it to initial capital cost to give the total lifetime cost of the transmission line.

Although system stability, reliability, radio interference, corona loss, voltage profile, short circuit level, and environmental aspects are important technical parameters that should be considered in the design and cost calculation of transmission lines, for the purposes of clarity, these were not included in the investigation of cost.

CAPITAL COSTS

The capital cost of a transmission line includes a number of components: towers, conductors, insulation, right-of-way, substations, compensation, labor and construction, engineering, and administration.

Most extra-high-voltage (EHV) lines are carried on steel towers. The main factor governing the cost of a tower is the weight of the steel; this can be expressed as a function of conductor size (Jones 1972).

$$\text{Tower's cost} = f \, (\text{conductor size})^{1/2} \quad (mm^2). \tag{1}$$

The conductor cost is about 34 percent of the total capital cost of a transmission line excluding substation and compensation cost (Dansfors, Hammarlund, and Peixoto 1977). Since the cost of other components is related to the conductor size, the choice of conductor is very important in cost optimization.

The cost of insulators for EHV transmission lines is about 4 percent of capital cost; this can be expressed as a function of the size of the conductor.

$$\text{Cost of insulator} = f\,(\text{size of conductor})^{1/2} \quad (mm^2). \qquad (2)$$

The cost of the right-of-way is related to the width of tower which is itself related to the voltage level of the line; the cost of the substation is effectively independent of the length of the line and is considered an initial cost. The cost of labor and construction is about 42 percent of the capital cost of the transmission line. The cost of research, planning, engineering, and design of a transmission line, along with supervision during construction, is about 5 percent of the total capital cost.

The cost of transmission can be expressed as a function of voltage level E and conductor size A, by an empirical formula (Jones 1972).

$$L_{cc} = \alpha + \beta A + \tau E + \delta E A^{1/2} \qquad (3)$$

where α, β, τ, and δ are constant coefficients. A is the conductor size of transmission (mm^2), E is the line-to-line voltage (kV), and L_{cc} is the unit cost of a line in $/mile. The first term shows initial cost, the second term expresses the cost related to the conductor, the third term shows the effect of voltage level on line cost, and the fourth term shows the effect of the cost of those components that are related to voltage level and conductor size of a transmission line. The values of α, β, τ, and δ can be estimated from observed costs of different transmission lines with different voltage levels and conductor sizes.

OPERATING COSTS

The operating cost of a transmission line includes the following components: losses, annual maintenance, and personnel.

The cost of power losses in a transmission line consists of extra power capacity to cover the losses in a maximum load condition and the cost of energy consumed in line. The cost of energy consumed is obtained by multiplying power losses of line at rated power and a factor called loss factor (LF). The approximate value of LF can be given by the following formula:

$$LF = 0.2F + 0.8F^2 \qquad (4)$$

Appendix 4   213

where F is the load factor. The cost of power losses in a transmission line over its useful life is from about 35 to 45 percent of its capital cost on a levelized basis. The annual maintenance and personnel cost is usually given as a percent of the capital cost; it is about 1 percent of capital for transmission line and 2 percent of the capital cost for substations and terminals (Casson 1963).

Levelized (capital and operating) cost of a transmission line can be obtained by Casson's formula:

$$V_P = (1 + \tau_L * a) \frac{L_{cc}}{P} + C_{PL} + C_{gp} + (1 + \tau_T * a) \frac{L_{sc}}{P} \qquad (5)$$

where,

$$a = \frac{1-(1+i)^{-n}}{i} \quad \text{(present worth factor)}$$

V_P = capitalized unit cost ($/MW/mile)
i = interest rate (%)
n = life of transmission line (yrs)
τ_L = line maintenance cost (% of L_{cc})
P = rated power of transmission line (MW)
L_{cc} = unit capital cost of transmission line ($/mile)

$$C_{PL} = h * LF * C_E * a * \frac{\Delta P_L}{P}$$

C_{PL} = cost of power losses ($)
LF = loss factor
C_E = cost of energy at bus bar ($/MWh)
h = 8,760 hours (the number of hours in a year)
ΔP_L = power losses in unit length of line (MW/mile)

$$C_{gp} = C_{pr} \frac{\Delta P_L}{P}$$

C_{gp} = cost of power replacement of losses ($)
C_{pr} = unit cost of power plant ($/MW)
L_{sc} = unit cost of terminals and compensation ($/mile)
τ_T = terminal maintenance cost (% of L_{sc}).

This formula gives the capitalized specific cost of power losses, terminal cost, and cost of compensation.

The resistive power loss (ΔP_L) of a three-phase line is given by equation (6):

$$\Delta P_L = \frac{P^2}{n\sigma A E^2 \cos^2 \phi} \qquad (6)$$

where:

ΔP_L = resistive power loss (MW/mile)
σ = conductivity of each subconductor (mile/Ω-mm^2)
n = number of subconductors in each phase

A = conductor cross-section area of each subconductor (mm²)
E = line-to-line voltage (kV)
Cosφ = power factor.

If β_1 and β_2 and α_t are the cost of series capacitor per Mvar, cost of shunt reactor per Mvar, and cost of substation equipment per MVA respectively, the cost of terminal and compensation for unit length of line is given by equation (7):

$$L_{sc} = 2\alpha_t/L + 0.5 X_L \beta_1 \frac{P}{E^2 \cos^2\phi} + 0.9 \frac{E^2 \beta_2}{X_c} * \frac{1}{P} \qquad (7)$$

where,

L_{sc} = cost of terminal and compensation ($/mile)
X_c = capacitive reactance per phase
L = length of line (miles)
X_L = inductive reactance per phase.

If ΔP_L, L_{sc}, and L_{cc} are substituted in equation (5), the specific cost of a transmission line is obtained as a function of P, E, n, A and is given by equation (8):

$$V_p = f(P, E, n, A) \qquad (8)$$

where V_p, P, E, n, and A are as defined above. Equation (8) can be used to find optimal loading, optimal voltage, and optimal conductor size. The cost of transmission of one kilowatt-hour is given in equation (9):

$$CET = \frac{TC}{ES} = \frac{V_p * P * L}{10^3 * 8760 * P_r * F * a} \qquad (9)$$

where,

CET = cost of energy transmission ($/kWh)
TC = total cost
ES = total energy supplied
P_r = power at receiving end (MW)
F = load factor
P = power at the sending end (MW).

Figure A-4.1 shows a typical transmission of energy over 100 miles at different voltage levels with the same conductor size (Moallem 1985). The data used for the cost calculations are given in Table A-4.1.

COST OF A HIGH-VOLTAGE DIRECT CURRENT LINE

The cost data for alternating current (AC) lines are readily available. However, this is not the case for direct current (DC) transmission lines. Due to a lack of well-established standards for DC equipment, the cost of DC

Appendix 4   215

lines varies between projects and manufacturers. The cost of a DC line is obtained by using the approach set forth by Dansfors, Hammarlund, and Peixoto (1977). The cost of a DC line is calculated from the known cost of AC lines for the same power transmission capacity. The following relations are used to calculate the DC transmission line costs:

$$CC_{DC} = 0.66 * CA_{DC}/CA_{AC} * CC_{AC}$$

$$IC_{DC} = 1.3/1.5 * 2/3 * V_{DC}/V_{PH} * IC_{AC}$$

$$TC_{DC} = 0.7 * V_{DC}/V_{DO} * (CA_{DC}/CA_{AC})^{1/3} * TC_{AC}$$

where,

CC_{DC} is the DC conductor cost, CA_{DC} is the DC conductor area, CA_{AC} is the AC conductor area, and CC_{AC} is the AC conductor cost; IC_{DC} is the DC insulator cost, V_{DC} is the DC line voltage, V_{PH} is the AC phase voltage, and IC_{AC} is the AC insulator cost; and TC_{DC} is the DC tower cost, V_{DO} is the DC voltage obtained by minimum clearance, CA_{DC} is the DC conductor area, CA_{AC} is the AC conductor area, and TC_{AC} are the AC tower costs.

$$EC_{DC} = (LHC_{DC}/LHC_{AC})^{0.75} * EC_{AC}$$

$$AC_{DC} = 0.8(AC_{AC})$$

where,

EC_{DC} is the DC erection cost, LHC_{DC} is the DC line hardware cost, LHC_{AC} is the AC line hardware cost, EC_{AC} is the AC erection cost; and AC_{DC} is the DC line administrative cost and AC_{AC} is the AC line administrative cost.

The cost of terminal equipment for DC lines is much higher than the cost for AC lines. Terminal equipment for DC lines includes converters (valves), harmonic filters, var injectors, and transformers. Figure A-4.2 shows typical costs for a DC converter. The cost of converters (in dollars per kilowatt) decreases with increases in the rating of the DC line. Figure A-4.3 shows the cost/kilowatt-hour of a $\pm$ 400-kilovolt DC line and a 500-kilovolt AC line for transferring 750 megawatts of power. A cost of $45/kilowatt is used for the converter terminal. The break-even distance, where the cost of the DC line is equal to the cost of the AC line, is 330 miles for this case. The break-even distance depends on the rating of the transmission line. Figure A-4.4 shows the variation in break-even distance with the increase in the rating of the DC line.

Figure A-4.1
Cost of Transmission Line for 100 Miles at Different Voltage Levels

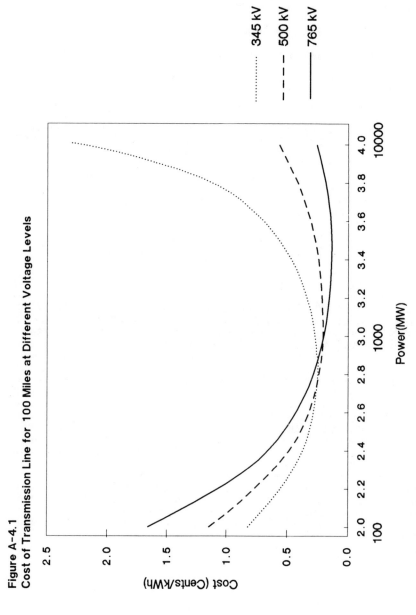

Source: Moallem, 1986.

216

Table A-4.1
Data Used for Transmission Line Cost

Capital cost of line including right-of-way, towers, and foundations (1985 dollars)

L_{cc} (345 kV) = $285,000/mile
L_{cc} (500 kV) = $420,000/mile
L_{cc} (765 kV) = $685,000/mile
These costs are for conductor size 1510 KCmil.

Cost of power replacement and bus bar energy:

Cost of power replacement = $1250/kW
Cost of energy on bus bar = 55.0 mills/kWh

Cost of AC terminal components:

Cost of transformer = $5000/MVA
Cost of series compensation = $1500/MVA
Cost of shunt compensation = $5500/MVA
Cost of circuit breaker = $5,000,000
(The cost of circuit breaker includes the cost of a three-phase circuit breaker with all measurement and relaying accessories)

Other parameters used in cost calculation:

Load factor (F) = 0.85
Loss factor (LF) = 0.75
Useful life of transmission line = 30 years
Rate of interest = 6%
Rate of inflation = 6%

Figure A-4.2
Typical Costs for a DC Converter

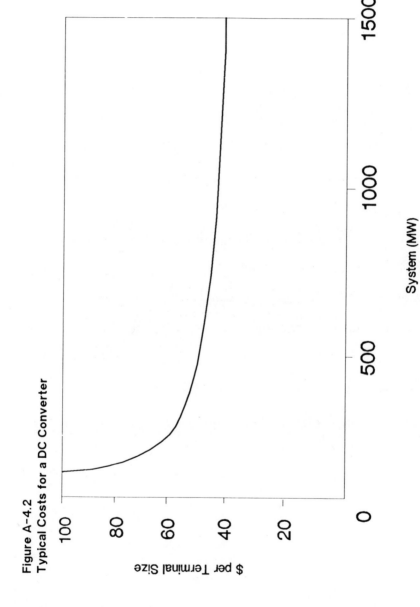

Source: Moallem, 1986.

Figure A-4.3
Break-Even Distance (Power=750 MW; Load Factor=0.85)

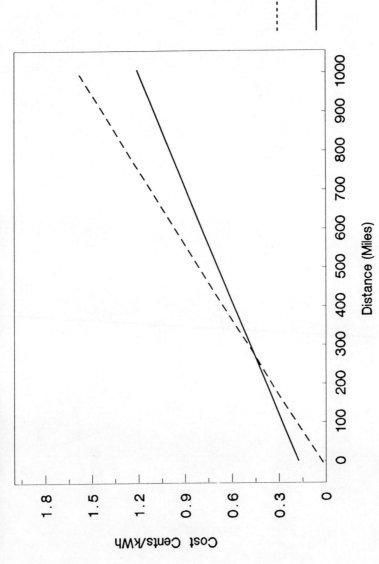

Source: Moallem, 1986.

Figure A-4.4
Variation in Break-Even Distance for AC and DC Transmission
Lines for Different Power Ratings

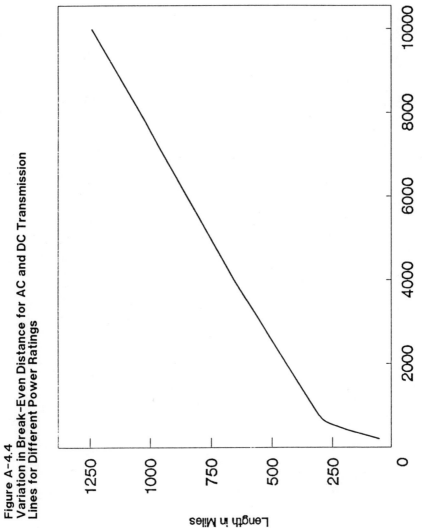

Source: Moallem, 1986.

Appendix 5:
Input Data Used in the
CONSOL Coal Research Model

Table A-5.1
Input Data Used in the CONSOL Coal Research Model

Variable	Data
General Assumptions	
Plant Sizes, Net:	300 MW, 500 MW, 750 MW
Plant Age:	New
Emission Criteria:	1979 NSPS
Particulate Removal:	ESP
FGD Process:	Wet limestone slurry with reheat
Sludge Disposal:	Forced oxidization and landfill
Sludge Disposal Cost:	$10/ton
Book Life:	30 years
Cost Factors	
Construction Cost Factors:	
West Virginia:	1.08
New Jersey:	1.18
Connecticut	
Financial Assumptions	
Cost of Capital:	6%
Cost of Debt:	6%
Debt Percentage:	50%
Income Tax Rate:	50%
Other Taxes and Insurance:	2%
Investment Tax Credit:	10% (first year of operation only)
Depreciation:	
Base Plant:	15 years
FGD:	5 years
Inflation and Escalation Rates	
GNP Deflator:	6%
Capital Equipment:	6%
Operating Labor:	6%
General Supplies:	6%
Rail Transportation:	6%
Coal:	6%

Sources: CONSOL Coal Research Power Costing Model, 1987
Steans Catalytic Corporation, 1984

Appendix 6:
Coal Analysis and Quality Parameters

Table A-6.1
Coal Analysis and Quality Parameters (Northern West Virginia Coal)

Variable	Data	Variable	Data
Proximate Analysis (weight %)		Ash Fusion Temperatures	
Moisture	6		
Volatile Matter	37.65	Reducing/Oxidizing (deg F)	
Fixed Carbon	46.85	I.D.	2053/2405
Ash	9.5	Softening	2250/2465
		H.T.	2305/2485
Ultimate Analysis (weight %)		Fluid	2415/2550
Hydrogen	5.08	Base/Acid Ratio	0.4
Carbon	75.32		
Nitrogen	1.41	Slagging	
Oxygen	5.33		
Sulfur	2.66	Index	1.06
Ash	10.11	Class	Medium
Chlorine	0.09	Coefficient	0
Higher Heating Value	13830	Fouling	
		Index	0.41
Ash Elements Analysis		Class	Medium
NA2O	1.02	Coefficient	0
K2O	1.26		
CAO	5.54	Heat Specific Rates (lb/MMBtu)	
MGO	0.98		
FE2O3	17.7	Coal	76.92
TiO2	0.99	Moisture	4.62
P2O5	0.24	Sulfur	1.92
SiO2	44.35	Ash	7.31
AL2O3	21.25	Stoic Air (lb/lb)	10.23
SO3	5.66		
Total	98.99	Grindability	55

Source: CONSOL Coal Research Power Costing Model 1987

Appendix 7:
Power Plant Performance Parameters

Table A-7.1
Power Plant Performance Parameters (Northern West Virginia Coal)

Variable	Plant Size 300 MW	500 MW	750 MW
Auxiliary Power Requirements (kW)			
Steam Generation	3130	5215	7823
Pulverizers	1419	2365	3548
Air and Flue Gas	1341	2235	3353
Coal Handling	1079	1798	2698
Particulate Removal	727	1211	1817
Plant Solid Waste	94	156	235
Water and Liquid Waste	502	837	1256
Balance of Plant	8410	14014	21018
Flue Gas Desulfurization	4430	7212	10884
Total	21133	35145	52629
Gross Generation (kW)	321133	535145	802629
Turbine Heat Rate (Btu/kWh)	7765	7765	7765
Flue Gas Reheat Duty (MMBtu/hr)	30	51	77
Total Boiler Output (MMBtu/hr)	2523	4206	6310
Boiler Efficiency (%)	88.57	88.59	88.6
Coal Input (MMBtu/hr)	2852	4752	7128
Net Plant Heat Rate (Btu/kWh)	9505	9504	9504
Coal Input (tons/hr)	3047	5078	7616
Excess Air at Economizer (%)	109.67	182.77	274.15
Air Preheater Leak (%)	21	21	21
Flue Gas Analysis (ex A.P.H.)	8	8	8
Flow Rate (MACFM)	944	1573	2359
Temperature (deg F)	291	291	291
Acid Dew Point (deg F)	281	281	281
Fly Ash (ton/hr)	7.51	12.51	18.77
Total Ash (ton/hr)	10.73	17.87	26.81
Limestone (ton/hr)	8.24	13.73	20.59
Sludge (ton/hr)	25.96	43.26	64.89
Design Parameters			
Pulverizers (#)	3	4	6
Pulverizer Capacity	67.11	83.88	83.87
Furnace			
Input Plan (MBtu/ft$_2$/hr)	1808	1808	1808
Release Rate (MBtu/ft$_2$/hr)	127	127	127
Liberation Rate (Btu/ft$_3$)	19867	19867	19867
Plan Area (ft$_2$)	1686	2809	4214
Projected Surface (ft$_2$)	23990	39980	59967
Volume (ft$_3$)	143531	239198	358781
Conveyor Pass Velocity			
Design (ft/sec)	65	65	65
Maximum (ft/sec)	66	66	66
Particulate Removal			
Efficiency (%)	99.43	99.43	99.43
Ash Resistivity (ohm/cm)	1.34E+09	1.34E+09	1.34E+09
SCA (ft$_2$/MACFM)	308	308	308
Area (Mft$_2$)	291	484	727
Flue Gas Desulfurization			
Efficiency (%)	90	90	90
Sulfur Removal (%)	84.4	84.4	84.4

Source: CONSOL Coal Research Power Costing Model 1987

Appendix 8:
Constant Dollar Results for Comparative Cost Analysis

Table A-8.1
Comparative Cost Results for a New Plant in New Jersey
vs. Power from West Virginia (mills/kWh constant 1985 dollars)

Plant Location	Plant Size (MW)	Capacity Factor (%)	New Plant Capitol Cost	Energy Cost	Scrubber Cost	Bus Bar Cost[a]	Transmission Cost Scenarios[b]		Delivered Cost Scenarios[b]	
							Low	High	Low	High
New Jersey	300	65	6.95	16.14	4.90	31.70	---	---	31.70	
		80	5.65	16.14	4.22	29.07	---	---	29.07	
West Virginia	300	65	6.27	11.33	4.81	25.94	8.73	23.53	34.67	49.47
		80	5.09	11.33	4.15	23.49	8.71	23.52	32.20	47.01
New Jersey	500	65	5.88	16.14	4.24	29.31	---	---	29.31	
		80	4.78	16.14	3.69	27.13	---	---	27.13	
West Virginia	500	65	5.28	11.33	4.16	23.69	5.54	14.45	29.23	38.14
		80	4.29	11.33	3.62	21.66	5.52	14.43	27.10	36.09
New Jersey	750	65	5.19	16.14	4.17	28.10	---	---	28.10	
		80	4.22	16.14	3.63	26.15	---	---	26.15	
West Virginia	750	65	4.66	11.33	4.09	22.56	4.05	10.02	26.61	32.50
		80	3.78	11.33	3.56	20.74	4.02	9.99	24.76	30.73

	N.WV-NJ	N.WV-WV	
Coal Transportation Cost:	$18.85/ton	$5.77/ton	N.WV Coal Price: $25/ton

Inflation: 6%   Cost of Debt: 6%   Cost of Capital: 6%

		WV	NJ	Construction Schedules:		
Power Plant Capital Costs:	300MW	$963/kW	$1050/kW		WV	NJ
	500MW	$799/kW	$ 871/kW	Land Purchase	1/1987	1/1987
	750MW	$711/kW	$ 775/kW	Startup	6/1994	6/1994

[a]Total includes plant consumables and operation and maintenance.

Transmission right-of-way costs (500-kV line): Low: $506,000/mile; High: $1,520,320/mile.

Source: CONSOL Coal Research Power Costing/Coal Quality Model.
[b]Author' calculations.

Table A-8.2
Comparative Cost Results for a New Plant in Connecticut vs. Power from West Virginia
(mills/kWh constant 1985 dollars)

Plant Location	Plant Size (MW)	Capacity Factor (%)	New Plant Capitol Cost	Energy Cost	Scrubber Cost	Bus Bar Cost[a]	Transmission Cost Scenarios[b]		Delivered Cost Scenarios[b]	
							Low	High	Low	High
Connecticut	300	65	6.53	16.57	4.80	31.49	---	---	31.49	
		80	5.31	16.57	4.15	28.99	---	---	28.99	
West Virginia	300	65	6.27	11.33	4.81	25.94	12.68	34.95	38.62	60.89
		80	5.09	11.33	4.15	23.49	12.66	34.93	36.15	58.42
Connecticut	500	65	5.57	16.57	4.16	29.26	---	---	29.26	
		80	4.52	16.57	3.62	27.19	---	---	27.19	
West Virginia	500	65	5.28	11.33	4.16	23.69	8.06	21.50	31.75	45.19
		80	4.29	11.33	3.62	21.66	8.03	21.47	29.69	43.13
Connecticut	750	65	4.95	16.57	4.08	28.15	---	---	28.15	
		80	4.02	16.57	3.56	26.28	---	---	26.28	
West Virginia	750	65	4.66	11.33	4.09	22.56	5.91	14.94	28.47	37.50
		80	3.78	11.33	3.56	20.74	5.86	14.89	26.60	35.63

	N.WV-CT	N.WV-WV	
Coal Transportation Cost:	$20.00/ton	$5.77/ton	N.WV Coal Price: $25/ton

Inflation: 6%   Cost of Debt: 6%   Cost of Capital: 6%

		WV	CT	Construction Schedules:		
					WV	CT
Power Plant Capital Costs:	300MW	$963/kW	$956/kW			
	500MW	$799/kW	$794/kW	Land Purchase	1/1987	1/1987
	750MW	$711/kW	$708/kW	Startup	6/1994	6/1994

[a]Total includes plant consumables and operation and maintenance.

Transmission right-of-way costs (500-kV line): Low: $506,000/mile; High: $1,520,320/mile.

Source: CONSOL Coal Research Power Costing/Coal Quality Model.
[b]Author' calculations.

Table A-8.3
Comparative Cost Results for a New Plant in New Jersey vs. Power from West Virginia:
   Two-Year Accelerated Construction Schedule for West Virginia Plant
   (mills/kWh constant 1985 dollars)

Plant Location	Plant Size (MW)	Capacity Factor (%)	New Plant Capitol Cost	Energy Cost	Scrubber Cost	Bus Bar Cost[a]	Transmission Cost Scenarios[b]		Delivered Cost Scenarios[b]	
							Low	High	Low	High
New Jersey	300	65	6.95	16.14	4.90	31.70	---	---	31.70	
		80	5.65	16.14	4.22	29.07	---	---	29.07	
West Virginia	300	65	6.18	11.33	4.79	25.83	8.73	23.53	34.56	49.36
		80	5.02	11.33	4.13	23.40	8.71	23.52	32.11	46.92
New Jersey	500	65	5.88	16.14	4.24	29.31	---	---	29.31	
		80	4.78	16.14	3.69	27.13	---	---	27.13	
West Virginia	500	65	5.21	11.33	4.14	23.60	5.54	14.45	29.14	38.05
		80	4.23	11.33	3.61	21.59	5.52	14.43	27.11	36.02
New Jersey	750	65	5.19	16.14	4.17	28.10	---	---	28.10	
		80	4.22	16.14	3.63	26.15	---	---	26.15	
West Virginia	750	65	4.59	11.33	4.07	22.48	4.04	10.02	26.52	32.50
		80	3.73	11.33	3.55	20.68	4.02	10.00	24.70	30.68

	N.WV-NJ	N.WV-WV	
Coal Transportation Cost:	$18.85/ton	$5.77/ton	N.WV Coal Price: $25/ton

Inflation: 6%   Cost of Debt: 6%   Cost of Capital: 6%

		WV	NJ	Construction Schedules:		
					WV	NJ
Power Plant Capital Costs:	300MW	$963/kW	$1050/kW	Land Purchase	1/1987	1/1987
	500MW	$799/kW	$ 871/kW	Startup	6/1992	6/1994
	750MW	$711/kW	$ 775/kW			

[a] Total includes plant consumables and operation and maintenance.

Transmission right-of-way costs (500-kV line): Low: $506,000/mile; High: $1,520,320/mile.

Source: CONSOL Coal Research Power Costing/Coal Quality Model.
[b] Author' calculations.

Table A-8.4

Comparative Cost Results for a New Plant in Connecticut vs. Power from West Virginia:
Two-Year Accelerated Construction Schedule for West Virginia Plant
(mills/kWh constant 1985 dollars)

Plant Location	Plant Size (MW)	Capacity Factor (%)	New Plant Capitol Cost	Energy Cost	Scrubber Cost	Bus Bar Cost[a]	Transmission Cost Scenarios[b] Low	Transmission Cost Scenarios[b] High	Delivered Cost Scenarios[b] Low	Delivered Cost Scenarios[b] High
Connecticut	300	65	6.53	16.57	4.80	31.49	---	---	31.49	
		80	5.31	16.57	4.15	28.99	---	---	28.99	
West Virginia	300	65	6.18	11.33	4.79	25.83	12.68	34.95	38.51	60.78
		80	5.02	11.33	4.13	23.40	12.66	34.93	36.06	58.33
Connecticut	500	65	5.57	16.57	4.16	29.26	---	---	29.26	
		80	4.52	16.57	3.62	27.19	---	---	27.19	
West Virginia	500	65	5.21	11.33	4.14	23.60	8.06	21.50	31.66	45.10
		80	4.23	11.33	3.61	21.59	8.03	21.47	29.62	43.06
Connecticut	750	65	4.95	16.57	4.08	28.15	---	---	28.15	
		80	4.02	16.57	3.56	26.28	---	---	26.28	
West Virginia	750	65	4.59	11.33	4.07	22.48	5.90	14.93	28.38	37.41
		80	3.73	11.33	3.55	20.68	5.86	14.89	26.54	35.57

	N.WV-CT	N.WV-WV	
Coal Transportation Cost:	$20.00/ton	$5.77/ton	N.WV Coal Price: $25/ton

Inflation: 6%   Cost of Debt: 6%   Cost of Capital: 6%

		WV	CT	Construction Schedules:	WV	CT
Power Plant Capital Costs:	300MW	$963/kW	$956/kW			
	500MW	$799/kW	$794/kW	Land Purchase	1/1987	1/1987
	750MW	$711/kW	$708/kW	Startup	6/1992	6/1994

[a]Total includes plant consumables and operation and maintenance.

Transmission right-of-way costs (500-kV line): Low: $506,000/mile; High: $1,520,320/mile.

Source: CONSOL Coal Research Power Costing/Coal Quality Model.
[b]Author' calculations.

Table A-8.5
Comparative Cost Results for a New Plant in New Jersey vs. Power from West Virginia:
  Two-Year Accelerated Construction Schedule for West Virginia Plant, Two-Year Delayed Schedule
  for New Jersey Plant
(mills/kWh constant 1985 dollars)

Plant Location	Plant Size (MW)	Capacity Factor (%)	New Plant Capitol Cost	Energy Cost	Scrubber Cost	Bus Bar Cost[a]	Transmission Cost Scenarios[b]		Delivered Cost Scenarios[b]	
							Low	High	Low	High
New Jersey	300	65	6.95	16.14	4.90	31.70	---	---	31.70	
		80	5.65	16.14	4.22	29.07	---	---	29.07	
West Virginia	300	65	6.18	11.33	4.79	25.83	8.73	23.53	34.56	49.36
		80	5.02	11.33	4.13	23.40	8.71	23.52	32.11	46.92
New Jersey	500	65	5.88	16.14	4.24	29.31	---	---	29.31	
		80	4.78	16.14	3.69	27.13	---	---	27.13	
West Virginia	500	65	5.21	11.33	4.14	23.60	5.54	14.45	29.14	38.05
		80	4.23	11.33	3.61	21.59	5.52	14.43	27.11	36.02
New Jersey	750	65	5.19	16.14	4.17	28.10	---	---	28.10	
		80	4.22	16.14	3.63	26.15	---	---	26.15	
West Virginia	750	65	4.59	11.33	4.07	22.48	4.04	10.02	26.52	32.50
		80	3.73	11.33	3.55	20.68	4.02	10.00	24.70	30.68

Coal Transportation Cost: N.WV-NJ $18.85/ton   N.WV-WV $5.77/ton   N.WV Coal Price: $25/ton

Inflation: 6%   Cost of Debt: 6%   Cost of Capital: 6%

	WV	NJ	Construction Schedules:	WV	NJ
Power Plant Capital Costs: 300MW	$963/kW	$1050/kW			
500MW	$799/kW	$ 871/kW	Land Purchase	1/1987	1/1987
750MW	$711/kW	$ 775/kW	Startup	6/1992	6/1996

[a]Total includes plant consumables and operation and maintenance.

Transmission right-of-way costs (500-kV line): Low: $506,000/mile; High: $1,520,320/mile.

Source: CONSOL Coal Research Power Costing/Coal Quality Model.
[b]Author' calculations.

Table A-8.6

Comparative Cost Results for a New Plant in Connecticut vs. Power from West Virginia:
Two-Year Accelerated Construction Schedule for West Virginia Plant, Two-Year Delayed Schedule for Connecticut Plant
(mills/kWh constant 1985 dollars)

Plant Location	Plant Size (MW)	Capacity Factor (%)	New Plant Capitol Cost	Energy Cost	Scrubber Cost	Bus Bar Cost[a]	Transmission Cost Scenarios[b] Low	Transmission Cost Scenarios[b] High	Delivered Cost Scenarios[b] Low	Delivered Cost Scenarios[b] High
Connecticut	300	65	6.53	16.57	4.80	31.49	---	---	133.47	
		80	5.31	16.57	4.15	28.99	---	---	122.90	
West Virginia	300	65	6.18	11.33	4.79	25.83	12.68	34.95	38.51	60.78
		80	5.02	11.33	4.13	23.40	12.66	34.93	36.06	58.33
Connecticut	500	65	5.57	16.57	4.16	29.26	---	---	124.05	
		80	19.17	16.57	3.62	27.19	---	---	115.24	
West Virginia	500	65	5.21	11.33	4.14	23.60	8.06	8.06	31.66	45.10
		80	4.23	11.33	3.61	21.59	8.03	8.03	29.62	43.06
Connecticut	750	65	4.95	16.57	4.08	28.15	---	---	119.32	
		80	4.02	16.57	3.56	26.28	---	---	111.40	
West Virginia	750	65	4.59	11.33	4.07	22.48	5.90	5.90	28.38	37.41
		80	3.73	11.33	3.55	20.68	5.86	5.86	26.54	35.57

	N.WV-CT	N.WV-WV	
Coal Transportation Cost:	$20.00/ton	$5.77/ton	N.WV Coal Price: $25/ton

Inflation: 6%   Cost of Debt: 6%   Cost of Capital: 6%

		WV	CT	Construction Schedules:		
Power Plant Capital Costs:	300MW	$963/kW	$956/kW		WV	CT
	500MW	$799/kW	$794/kW	Land Purchase	1/1987	1/1987
	750MW	$711/kW	$708/kW	Startup	6/1992	6/1996

[a]Total includes plant consumables and operation and maintenance.

Transmission right-of-way costs (500-kV line): Low: $506,000/mile; High: $1,520,320/mile.

Source: CONSOL Coal Research Power Costing/Coal Quality Model.
[b]Author' calculations.

Index

Appalachians, 20, 26
  acid rain: legislation, 20, 67, 136-139; control, 5,9; impacts, 132-134
Allegheny Power System, 87
Appalachian Power Company, 33
American Electric Power Company, 6, 75, 87
Appalachian Project, 62
Atomic Energy Act, 66

Burns and Roe Company, 47
Bulk power trade, 85-90

cogeneration, 68-72
coal gasification, 62
Canada: acid rain control, 139-142; power exports, 5, 19-20, 91-94, 141; electricity demand, 39; utilities, 11, 75-77
Capacity power, 4, 90-94
Center for Clean Air Policy, 5, 139, 174-176
Certificate of convenience and necessity, 11
Connecticut, 5
Clean Air Act, 16-17, 19, 24, 67, 132, 134-136, 149-150
Cool Water IGCC plant, 55, 62
Comparative Cost Analysis, 179-193
Coal: reserves, 60; transportation, 183
Control areas, 75-77, 85-87

Citizens for the Protection of Washington County (CPWC) 160-162
CONSOL Coal Research Model 179-180

Direct Current (DC) transmission, 81-84
District Heating, 68-69

Economy energy, 4, 85-90, 95
Edison, Thomas A., 9-10
East Central Area Reliability Council (ECAR), 40-49, 95
Electrostatic precipitators, 58
Economic development rates, 125-126
Environmental Impact Statements 148, 151-152
Electricity prices, effects 120-125

Fluidized-bed combustion, 61
Federal Energy Regulatory Commission (FERC), 14-15, 69-70, 75, 87
Fly ash, 58
Federal Emergency Management Agency (FEMA) 65

General Public Utilities, 94

Integrated coal-gasification combined cycle (IGCC), 55-56, 61-62

John E. Amos power plant, 163

Kaiparowits power plant, 151

Limerick II power plant, 19
Long Island Lighting Company, 66

Moore, Arch A., 5-6
Mid-Atlantic Area Reliability Council (MAAC), 27, 30-49, 95
Maryland, 159-162
Mountaineer power plant, 150-153, 162-164
Maryland Power Plant Siting Program, 159-162

National Governors Association Electricity Transmission Task Force, 6, 97, 153, 155-156
New England, 42, 58, 71-72
New England Power Pool (NEPOOL), 42-49, 174-177
New Hampshire, 5-6
North American Electric Reliability Council (NERC), 30-49, 86
New Source Performance Standards (NSPS), 24, 26
Nitrogen Oxide, 24, 61
Nuclear power plant evacuation planning, 65-66
New England Electric System, 176-178
Ontario, acid rain control, 141

Pennsylvania, 155-156
Public Utilities Regulatory Policy Act (PURPA), 69-70
Power Plant and Industrial Fuels Use Act, 16, 26, 57

Pulverized-coal combustion, 58, 61, 182
Power plant: outages, 49, 65-66; efficiencies, 58, 61-62; retrofitting, 66-67
Power pools, 86-88
Pennsylvania-Maryland-New Jersey Interconnection (PJM), 87
Power wheeling, 87-88
Power plant siting: permits, 148-151; public opposition, 158-169; methods, 159-162
Power plant costing models, 177-179

Requirements customers, 14
Reserve margins, 40-47

Sulfur dioxide, 24-26, 61-62
Scrubbers, 24-26, 58, 61, 67, 137, 149
Shoreham power plant, 65-66
Seabrook power plant, 65

Three Mile Island power plant, 64-65
Transmission line: costs, 183-186; extra high voltage lines, 77; historical development, 75-77; power losses, 77; reactive power, 71, 79-82; stability, 79; superconductivity, 84; siting, 153-158; health effects, 153-155

Virginia-Carolina subregion (VACAR), 42-48

West Virginia, 4-7, 33, 61-62, 179-193
West Virginia Public Energy Authority, 5-6, 151
Wharton Econometric Forecasting Associates, 36-38
West Virginia input-output model, 101-110